8

ENDOCRINOLOGY RESEARCH AND CLINICAL DEVELOPMENTS

GLUCAGON

STRUCTURE, BIOSYNTHESIS AND PHYSIOLOGICAL EFFECTS

ENDOCRINOLOGY RESEARCH AND CLINICAL DEVELOPMENTS

Additional books in this series can be found on Nova's website under the Series tab.

Additional e-books in this series can be found on Nova's website under the e-book tab.

ENDOCRINOLOGY RESEARCH AND CLINICAL DEVELOPMENTS

GLUCAGON

STRUCTURE, BIOSYNTHESIS AND PHYSIOLOGICAL EFFECTS

AHMED AL-JEBAWI
EDITOR

New York

For permission to use material from this book please contact us:
Telephone 631-231-7269; Fax 631-231-8175
Web Site: http://www.novapublishers.com

Library of Congress Cataloging-in-Publication Data

ISBN: 978-1-62618-147-2
Library of Congress Control Number: 2013931734

Published by Nova Science Publishers, Inc. † New York

Contents

Preface

As Diabetes Mellitus is becoming alarmingly a common disease worldwide, researchers are trying to find more effective ways of treating it and controlling its complications. Despite the vast knowledge that we gained on glucagon since its discovery, treating people with diabetes remained limited to replacing Insulin, stimulating its secretion or improving its action for many years. Targeting glucagon pathway was not thought as an a valuable approach to treating diabetes until the first incretin mimetic was proved to be highly effective for not only stimulating the secretion of insulin but also for suppressing the glucagon secretion. The concept of bi-hormonal abnormality has changed the way we approach and treat diabetes. The fact that diabetic people do not respond to the anti-hyperglycemic treatment in the same way clearly indicates that our prescribing habits should be based on their individual needs.

This is the first edition of "Glucagon: Structure, Biosynthesis and Physiological Effects". The book was designed to answer important questions about glucagon including those related to its normal function as well as its role in the pathogenesis of diabetes. It also summarizes recent findings regarding the potential clinical use of different strategies for correcting glucagon disorders among people suffering from diabetes. It provides readers with a greater understanding of Glucagon agonist/ antagonist therapy. Such therapy may not only provide treatment for glucose metabolic disorders but may also revolutionize the treatment of various other diseases including cardiovascular diseases, obesity and fatty liver disease.

The readers to whom this book is addressed are those who want to have a better understanding of the rapidly progressing advances in glucagon research. All endocrinologists, internists, primary care physicians, physiologists, biochemists and medical students should find this book a good reference with all the essential information they need. Without going through the molecular science details, this book provides a basic, conceptual-level description of the main topics related to glucagon structure, biosynthesis and physiological effects. It integrates both basic and clinical sciences in such a way that readers will feel the clinical applications of the information they read.

In: Glucagon
Editor: Ahmed Al-Jebawi

ISBN: 978-1-62618-147-2

Chapter 1

History of Glucagon and GLP-1

Alaaeldin Bashier* and Puja Murli Thadani
Endocrine Division-Medical Department, Dubai Hospital, Dubai, UAE

Abstract

Research in the field of gut hormones has started as early as 1888, with the work of Pavlov, who tried to prove that most of gut functions are controlled by the nervous system. In 1902 Bayliss and Starling discovered the chemical messenger that was later called the (secretin) hormone. This was followed by the first human experience in treating diabetes with duodenal extract by Moore et al. in 1906. Many years passed and there was no definite answer to the relation of diabetes to this pancreatic extract. In 1939/1940 Leow and coworkers published a paper that concluded no relation between diabetes and secretin. This conclusion resulted in lack of interest in research, which continued for almost three decades. The discovery of radioimmunoassay for insulin and glucagon has propelled the research in this field towards incretin based therapies. In 1971 the chemical structure of the first incretin was deduced by Brown and Dryburgh and was given the name gastric inhibitory peptide (GIP). In 1985 the second incretin glucagon like peptide 1 (GLP-1) was discovered. FDA has approved the first incretin-based therapy exenatide for treatment of type2 diabetes in 2005, followed one year later (2006) by Sitagliptin. After the discovery of insulin the scientists' noticed a short period of hyperglycemia before the desired effect of insulin, this was

* Corresponding author: Alaaeldin11@gmail.com; P.O.Box 94132.

found to be due to glucagon, which was discovered in 1923. In 1956 Bromer et al. managed to purify and sequence glucagon, this was followed few years later by development of the radioimmunoassay for glucagon and insulin 1959 and 1960. Unger et al. have showed the role of glucagon in the development of hyperglycemia in 1975. The discovery of glucagon gene and glucagon receptor has helped the scientists to keep searching for new strategies to target glucagon receptor and pathways. This chapter will discuss in detail different milestones and landmark research related to glucagon and incretin hormones.

List of Terms

GIP	Gastric inhibitory peptide
GLP-1	Glucagon like peptide 1
VIP	Vasoactive intestinal peptide
GRF	Growth hormone releasing factor
GLP-2	Glucagon like peptide 2
RIA	Radioimmunoassay
AMP	Adenosine monophosphate
GIT	Gastro intestinal tract
HCL	Hydrochloric acid
DPP-4	Di peptidyl peptidase 4
FDA	Food and drug administration
cAMP	Cyclic adenosine monophaosphate

Introduction

"Great discoveries of Science are those which can be seen sometimes long afterwards – to have changed our way of thought about a natural phenomenon and so to have turned the course of discovery in a new fruitful direction which none has earlier foreseen".

R.A. Gregory

Glucagon is a 29 amino acid peptide hormone that plays an important role in glucose homeostasis. It has a molecular weight of 3485 daltons, and is secreted from α cells of the islets of Langerhans following posttranslational processing of a larger precursor, proglucagon. This hormone is a member of a superfamily of polypeptides that includes secretin (Mutt et al. 1970), gastric inhibitory peptide (GIP) (Brown and Dryburgh 1971), vasoactive intestinal peptide (VIP) (Mutt and Said 1974), pre-albumin (Jornvall et al. 1981), and a growth hormone-releasing factor (hpGRF) (Spiess et al. 1982).

The glucagon gene is a single copy gene that has been localized to band q36-37 on human chromosome 2 (Tricoli et al. 1984). It encodes for proglucagon which is a tissue specific peptide that produces different molecules on posttranslational processing depending on the tissue site. In the brain and alpha cell it produces glucagon, while in the intestines it is converted into glicentin, oxyntomodulin, GLP-1, and GLP-2. The posttranslational processing of proglucagon in the islet α cells is influenced by prohormone convertase-2 enzyme (PC2) (Rouillé Y 1994).

As glucagon is secreted from alpha cells, GLP-1 and GLP-2 are secreted from L cells of large intestines, also called entero-endocrine cells. GLP-1 stimulates insulin secretion and inhibits glucagon release from alpha cells, delays gastric emptying, promotes satiety, and may, through unknown mechanisms, regulate glycogen synthesis in adipose tissues and muscles (Drucker D J 1998). The secretion of glucagon, as the name implies "a hormone of energy need", increases rapidly in cases of hypoglycemia. In patients with diabetes this counter-regulatory response is impaired and defective predisposing them to hypoglycemia.

Milestones in Glucagon Development and Research

Banting and Best, who discovered Insulin in 1921, noticed during their studies a mild and reproducible increase in blood glucose before the expected hypoglycemic effect of the pancreatic extracts, for which they could not find an explanation. Subsequently, in 1923, Murlin suggested the presence of a gluconeogenic substance in the pancreatic extracts, which he called Glucagon meaning "the mobilizer of glucose" (Murlin et al. 1923).

In the following years researchers continued to study and understand the physiological relevance of glucagon using non-purified pancreatic extracts until 1956. During this year Bromer not only managed to review and modify the various steps involved in the isolation of and purification of glucagon, but also successfully established its amino acid sequence. He found an identical sequence in humans, cattle and pigs. (Bromer et al. 1956, 1983). Further progress was achieved by Blundell *et al.*, (1983) and Wunsch *et al.*, 1972) in determining the structure of the molecule and synthesis of the hormone. These observations led to more progress in glucagon-related research enabling determination of the structure of the molecule and understanding the processes involved in its synthesis.

In 1959, Unger *et al.*, described the development of a radioimmunoassay (RIA) for glucagon in the paper titled "Glucagon antibodies and their use for immunoassay for glucagon" (Unger *et al.*, 1959). This work was based on the research carried out by Yalow and Berson on the development of a RIA for insulin. Although the RIA for glucagon was developed after that for insulin, the work of Unger et al. was published one year ahead of Yalow's work. Dr. Roger Unger has acknowledged this in his following work (Unger *et al.*, 1973), where he clearly stated, "Glucagon was the second hormone to be measured by means of the radio-immuno-chemical techniques of Berson and Yalow". Thus, the work of Yalow and Berson, "Immunoassay of endogenous insulin in man" is undisputedly considered to be the earliest one describing the first RIA of a peptide protein hormone, and is looked upon as a critical point in endocrinology research. (Yalow and Berson 1960) (Lefebvre, 2011).

Although glucagon was discovered in 1923 it has not been linked to diabetes until the 1975 with Unger's "Bi-hormonal theory". The theory in Unger's own words: "The uni-hormonal deficiency hypotheses of diabetes mellitus, in which all of the metabolic derangements of the disease are attributed to absolute or relative insulin lack, originated with the classic experiments of von Mering and Minkowski in 18891 and was reinforced by those of Banting and Best in 1922. However, developments in glucagon research raise the possibility that diabetic abnormalities in glucose homeostasis are the consequences of a bi-hormonal disorder in which a relative or absolute deficiency of insulin and a relative or absolute excess of glucagon both are etiological factors. According to this hypothesis, relative or absolute hypo-secretion of insulin would account primarily for underutilization, while relative hyper-secretion of glucagon would be the major factor in the over production of glucose at a rate which exceeded its utilization, resulting in endogenous hyperglycemia." (Unger et al. 1975)

Unger based his hypothesis on the following facts. First, endogenous hyperglycemia has never been seen in the absence of glucagon. This does not only include spontaneous diabetes in humans and diabetes in experimental animals but also hyperglycemia associated with burns, myocardial infarction … etc. secondly, somatostatin which is known to suppress both insulin and glucagon release, does not cause hyperglycemia unless glucagon levels were restored by exogenous glucagon. Thirdly, there is evidence that somatostatin-induced reduction in glucagon levels is associated with a reduction in blood glucose levels (Unger et al. 1975).

Contradictory to the previous belief that glucagon acts synergistically promoting both glucose mobilization in the liver and glucose utilization by the peripheral tissues, studies by Lefebvre et al., established glucagon as " a hormone of energy need" in circumstances of stress and hypoglycemia (Unger RH, Lefebvre PJ, 1972). This observation supports the current use of glucagon as a treatment for severe hypoglycemia. Historically Assan *et al.*, (1969) first described the defect in glucagon secretion in relation to diabetes. In their work they reported high glucagon levels in patients with diabetic ketoacidosis and its rapid regression with treatment (Assan *et al.*, 1969). Later the more universal hyperglucagonemia (either absolute or relative) in diabetic patients was described by (Unger *et al.*, 1970). The most recent observation in glucagon related research suggests that glucagon receptor knock out in mice prevents streptozocin-induced type1 diabetes. (Lee Y, 2011)

Despite the fact that the glucagon gene was discovered and located on chromosome 2 in 1984 (Tricoli *et al.*, 1984), White and Saunders (1986) identified the structure of the glucagon gene two years later. They found that the gene is approximately 9.4 kilobases in length and comprises six exons and five introns. In 1993 the glucagon receptor has been identified and cloned, through the work of Jelinek et al. In their study a complementary DNA clone for the glucagon receptor was isolated by an expression cloning strategy. An increase in intracellular cAMP was seen upon binding of the cloned receptor to glucagon. It also transduced a signal that led to an increased concentration of intracellular calcium. They also concluded that there are similarities between the glucagon receptor and calcitonin and parathyroid hormone receptors (Jelinek et al., 1993).

One of the breakthrough discoveries in medicine was the discovery of the G-proteins, which received a Nobel Prize in 1994. In his award ceremony speech, Professor Bertil B. Fredholm of the Nobel Committee at the Karolinska Institute said “About 25 years ago, Martin Rodbell and his colleagues decided to investigate how a chemical signal - a hormone - that came in contact with the outer surface of the cell membrane could bring about changes on the inside of the same membrane. They discovered that the transduction of signals across the cell membrane could be described as a three-step process. First the cell must recognize what kind of chemical signal is reaching it from other parts of the body - this requires what Rodbell called the discriminator. The last step in the signaling pathway is an amplifier, which ensures that the signal created inside the cell is strong enough to make something happen. The major breakthrough was Martin Rodbell's realization that there was a switch between these two steps, and that this switch, which he called the transducer, could be turned on by a high-energy compound, guanosine triphosphate. The letter G in G protein stands for guanosine triphosphate.

At this point, Alfred Gilman and his colleagues took over. Using a combination of genetic and biochemical techniques they managed, after a heroic effort, to isolate the G protein from all the other parts of the cell membrane. The workings of the protein could then be studied. Among other things, Gilman showed that the G protein works like a timed switch that allows the signal to go through just long enough. G proteins might perhaps be compared to those little gadgets that can be plugged into a telephone and that make it possible - with a phone call - to turn lamps on and off, start electric heaters, or draw curtains, depending entirely on what the gadget is connected to." (Nils Ringertz 1997, www.Nobelprize .org)

The Gilman and co-workers used lymphoma cells, as the mutated forms contain a normal receptor and a normal cyclic AMP-generating enzyme. The problem they faced was that these cells lack the transducer so it cannot be activated. They managed to isolate the G-protein from the brain and insert it into the mutated lymphoma cell; hence restoring it's function.

Incretins (GIP and GLP-1)

The discovery of incretins is one of the great discoveries as it has changed the way people think of GIT function. Historically Pavlov is considered to be one of the pioneers in gastro-intestinal research as he is credited with laying the foundation of modern gastroenterology. S.P. Botkin and his Doctrine of "Nervism", which stated that most bodily functions were regulated by the nervous system influenced Pavlov's early research (R. A. Gregory).

Trying to prove that GIT function is controlled by neural connections, Pavlov (1888) created a pancreatic fistula in a conscious dog and divided the vagus at the neck. After waiting for a few days for the cardiac fibers to degenerate, he stimulated the peripheral end of the vagus, and managed to obtain a free flow of pancreatic juice from the fistula. Subsequently, in 1889, he conducted another experiment where he created a gastric and esophageal fistula in a conscious dog. Through this experiment he found that prior vagotomy or administration of atropine prevented flow of gastric juice rich in acid mucus and pepsin. These two experiments carried out by Pavlov proved beyond doubt that vagal stimulation causes secretions from the stomach and pancreas (R. A. Gregory).

In 1894 Dolinski repeated a previous study by Leuret and Lassaigne (Mutt 1959) and found that introduction of dilute HCL into the duodenum resulted in the production of profuse and watery pancreatic juice through a pancreatic fistula. These observations shed light on a new hypothesis suggesting that the presence of hydrochloric acid in the duodenum acts as another stimulant of pancreatic secretion. Two years later, Pavlov, investigated the influence of this hypothesis on his previous observations. In the following experiment, he occluded the pyloric canal by a plug of cotton wool soaked with bicarbonate solution to prevent entry of acid to the duodenum. Then he stimulated the cervical vagus. His observations this time were different in comparison to his previous work. This experiment demonstrated a slow flow of viscid juice, which represented the actual effect of vagal stimulations (R. A. Gregory)

These intriguing findings encouraged further research to find an explanation for the differences in results. For example, Popielski (1901) showed that acid in the duodenum stimulated pancreatic secretions even after neural connections at the splanchnics, spinal cord, and celiac ganglia were sectioned in addition to destruction of the medulla of the spinal cord. On the other hand, Wertheimer and Le Page (1901) failed to stimulate pancreatic secretion in a similar denervation study after intravenous infusion of hydrochloric acid. At this point a local reflex pathway, which could be stimulated by fibers other than the vagus was suggested.

Unfortunately these experiments failed to identify the nature of a second local reflex pathway, which stimulates pancreatic secretions until 1902 when Bayliss and Sterling first introduced the concept of a chemical messenger, through their study in an anaesthetized dog. They tied a loop of jejunum at both ends and sectioned all nerve supply to that segment so that it was connected to the rest of the body by blood vessels only. A weak HCL was introduced to the duodenum and resulted in pancreatic secretion. They then extracted and filtered the jejunal mucosa and injected it intravenously. To their surprise this also resulted in pancreatic secretion, thereby leading to the conceptualization of a chemical messenger.

Bayliss and Starling later suggested that this substance, i.e. chemical messenger must have been secreted from the intestines into blood stream to stimulate pancreatic juice (Bayliss and Starling 1902) and they labeled it "secretin" (Wook Kim 2008). Two years later in their 'Cornian lectures' at the royal college of physicians they described the action of the chemical messenger and gave it the name hormone (Bayliss and Starling, 1904).

Additionally, Starling published the first notes on the relation of duodenal extracts with diabetes although the first experiment using upper GI extracts in the treatment of diabetes was carried out by Moore et al., (1906). This study aimed to test two hypotheses. One of them postulated that duodenal extracts stimulate pancreatic secretions, while the other proposed that glycosuria in patients with diabetes results from a functional disturbance due to the absence of the intestinal extract. The study was based on three patients with newly diagnosed juvenile diabetes and this group included one male patient aged 25 years, a 7-year old boy and a 9-year old girl. These patients were given acid extracts of porcine duodenal mucosa three times daily, which was administered orally.

All three patients showed good response to treatment demonstrated by the disappearance of glycosuria 3.5-12 weeks after initiation of treatment. Of note, one patient developed glycosuria two months after discontinuation of treatment with extracts of porcine duodenal mucosa. This patient is known to have died in the following months, the cause of death being attributed to pulmonary tuberculosis. In another study carried out by the same authors, patients treated with intestinal extracts experienced negative results. Moore et al., (1906) thereby concluded that no definitive deductions could be drawn from these studies, but their work formed the basis for future research.

Interestingly, failure to achieve positive results in the second study mentioned above can be explained by data available from current studies. It is now known that gut hormones are peptides that will be destroyed when administered orally and patients newly diagnosed with juvenile diabetes may enter into a honeymoon period, which is the most likely explanation for the improvement noted in the first three patients treated with extracts of porcine duodenal mucosa (Creutzfeldt et al., 2005).

In the following years, several animal studies were conducted to evaluate the effect of the hormone "secretin" identified earlier by Bayliss and Starling (1904). These studies were carried out using secretin (prepared according to the methods of secretin extraction) to assess its effect on fasting glucose and hyperglycaemia, which was induced by oral glucose intake. The mode of administration of the extract was varied using oral, intravenous, rectal and subcutaneous routes. The results obtained from these studies differed widely from one another (Creutzfeldt et al., 2005). In addition to this, La Barre (1939), who is credited with proposing the name "incretin", published a landmark paper with his co-worker Zunz in which he discussed ten relevant publications that explored the effect of intestinal extracts on blood glucose levels. The studies discussed in this paper also showed conflicting results, thus failing to precisely identify the effect of administered intestinal extracts on blood glucose levels (Zunz and La Barre, 1929).

These inconsistent results enthused many researchers to re-evaluate the effect of the intestinal extract on blood glucose. Loew and coworkers, Ivy and Gray, published the results of their controlled experiments in three papers that were considered to disprove the two major hypotheses on which incretin concepts were based. First, dilute hydrochloric acid stimulates duodenal mucosa and results in reduction of blood glucose, and second, injection or ingestion of the extract would give same effect (Loew et. al 1939) (Loew et al. 1940 (a, b)).

It is noteworthy that the discovery of a radioimmunoassay propelled research towards incretin-based therapy for treatment of diabetes after an interval of approximately 28years. McIntyre et al. (1964) showed that glucose given orally induces greater insulin responses compared to I.V glucose. This finding was also confirmed by two independent research groups (Elrick et al. 1964), and (Perley and Kipnis, 1967). Further advances were achieved in 1971, when JC Brown isolated and deduced the chemical structure of an incretin and named it "Gastric inhibitory peptide" (GIP) as it inhibited gastric acid secretion in dogs (Brown and Dryburgh, 1971). Another member of the incretin family, the glucagon like peptide-1 (GLP-1), was discovered in 1985. Importantly it was found to have potent insulinotropic properties and physiologically it was noted to be similar to GIP as its effect on insulin secretion is glucose dependent (Schmidt et al. 1985).

Nauck and co-workers further explored the effect of these newly diagnosed incretin hormones. They studied beta cell secretory response to oral glucose tolerance and iso-glycemic clamp study in 14 patients with type2 diabetes, and weight matched controls. In their experiment they noticed increased insulin response to oral glucose compared to intravenous administration and called this response "The incretin effect". Also, they noticed a reduced incretin effect in patients with type2 diabetes, despite normal levels of GIP suggesting loss of beta cell response to GIP in diabetic patients (Nauck, 1986). It is essential to note that they described GIP because it was the only measurable incretin hormone at that time. This study suggested that this diminished response could be explained either by hyposecretion or lack of effectiveness of another factor that has an incretin effect.

In the following year, Kreymann et al., (1987) confirmed this suggestion through their study of non-diabetic volunteers. They performed an oral glucose tolerance test and noticed that infusion of GLP-1, compared to normal saline, resulted in reduction of glucagon levels and markedly enhanced insulin secretion.

These two effects were reflected by reduction of peak plasma glucose. When compared to GIP infusion, GLP-1 was found to be a more effective stimulant of insulin release (Kreymann et al. 1987).

In 1992, Eng et al. reported the discovery of a new bioactive, 39 amino-acid peptide hormone called 'Exendin-4', which was isolated from the saliva of the glia monster. This hormone was found to have biological properties similar to the human GLP-1, which is a regulator of glucose metabolism and insulin secretion. Two years later, Deacon et al., (1994) described the role of the dipeptidyl-peptidase-4 (DPP-4) enzyme in the metabolism of GLP-1. They have also demonstrated that the biologically inactive N-terminally truncated metabolite, GLP-1-(9-36) is the major circulating peptide (Deacon et al., 1995).

The discovery of the DPP-4 enzyme was an important milestone as it facilitated the translation of research to clinical practice. Following this discovery much work has been done to identify the properties of these molecules and define their use clinically.

Further research in this direction continued until 2005 when the FDA as an additional treatment for type 2 Diabetes approved a synthetic version of Exedin-4, known as Exenatide.

Subsequently, the FDA for use approved Sitagliptin, which is a highly selective DPP-4 inhibitor, in 2006.

In a paper published in Diabetologia by Nauck et al. titled "Secretion of glucagon-like peptide-1 (GLP-1) in type 2 diabetes: what is up, what is down? ", they described their findings that GLP-1 is secreted from the L cells in the distal gut, while GIP is secreted from the K cells in the proximal small intestine (Nauck 2011)

Nobel Prize Awards for Glucagon Research

Glucagon research has been the royal way for Nobel Prize in physiology and medicine awards (Lefebvre 2011). Many of the pioneers in glucagon and gut hormones research have been awarded the prize.

Of note the work of Bayliss and starling has been declared in 1913 and 1914 to be worthy of a prize, but the war intervened, and when starling was nominated again the work was found to be very old to be eligible (Liljestrand, 1952).

Table 1.

Year	Land Mark Event
1902	Discovery of secretin
1903	Starling first suggestion of relation of duodenal extracts with diabetes
1904	Introduction of the term Hormone by Starling
1906	First experience in treating patients with diabetes with the duodenal extracts by Moore et al.
1923	Discovery of glucagon (Murlin et al.)
1923	Given the name glucagon
1929	Brown and Zunz paper on the effects of duodenal extract on the blood glucose
1939	Loew et al. papers that disproved the concept of incretin in reducing blood glucose
1956	Purification and sequencing of the glucagon (Bromer et al.)
1959	First description of a radioimmunoassay for glucagon (Unger et al.)
1964	Proving that oral glucose induces more insulin release compared to intravenous glucose by McIntyre et al. and Elrick et al.
1969	The defect first description of glucagon defect in diabetes (Assan et al.)
1971	Isolation and deduction of the chemical structure of an incretin (Brown and Dryburgh, 1971).
1971	Introduction of the name GIP (Brown et al.)
1984	Location of glucagon gene to chromosome 2
1985	Discovery of GLP-1
1986	Incretin effect (Nauck 1986)
1986	Identifying the structure of the glucagon gene (White and Saunders)
1993	Discovery of the glucagon receptor (Jelinek)
1994	Discovery of the DPPIV enzyme and it role in GLP-1 metabolism
2005	FDA approval of the first GLP-1 analogue or treatment of type2 diabetes
2006	Approval of the first DPP4 inhibitor for treatment of type2 diabetes

Pavlov's work, if taken into consideration as the earliest work in incretin-related research, would have been awarded the Nobel Prize in 1904. In his award-ceremony Pavlov stated "Essentially only one thing in life interests us: our psychical constitution, the mechanism of which was and is wrapped in darkness. All human resources, art, religion, literature, philosophy and historical sciences, all of them join in bringing light in this darkness. But man has still another powerful resource: natural science with its strictly objective methods. This science, as we all know, is making huge progress every day." He also sketched the broad lines on how can people successfully advance the future physiological research.

Earl Wilbur Sutherland was a biochemist whose work was mainly focused on determining the way in which hormones regulate body functions. In his early work he explained how adrenaline affects glucose regulation in the liver to give energy in cases of stress. Later, he discovered the cyclic adenosine 3'5'-monophosphate, or (cAMP), and showed that cAMP provides the link between hormones and intracellular processes. For his later work he was awarded Nobel Prize in physiology and medicine in 1971.

Sutherland together with de Duve hypothesized that pancreatic glucagon is produced from cells other than beta cells. De Duve managed to confirm this finding on his studies on guinea pigs. In these studies he showed that selectively damaging the islet alpha cells results in a drastic lowering of the glucagon content of the pancreas. Christian de Duve was later awarded Nobel Prize for his discovery of the lysosomes in (1974).

Finally the discovery of G-proteins and their role in signal transduction in cells by Martin Rodbell and Alfred Gilman was awarded a Nobel Prize in 1994. Martin Rodbell started the work on the G-proteins, which was completed later by Alfred Gilman, who used lymphoma cells, which possess a normal receptor and a normal cAMP-generating enzyme.

Conclusion

The discovery of the hormone secretin by Starling and Bayliss in 1902 established the basis for the incretin concept. This major breakthrough was followed by the discovery of glucagon in 1923. Due to failure to reach conclusive results on the relation of diabetes with these hormones, minimal progress was made in this field until 1960, when the radioimmunoassay for insulin and glucagon was discovered. This discovery had propelled the research further towards GLP-1 and other incretin based therapies. Although the incretin-related research started more than 100 years ago, the first clinical application of this research occurred in 2005 when exenatide was approved by FDA.

Since discovery of insulin in 1921, there was a uniform belief that all metabolic defects in diabetes are caused by insulin deficiency. In 1975, Unger could support his new hypothesis that suggests an essential role of glucagon in the pathogenesis of diabetes, with scientific evidence. This new concept opened the doors for a new era in research.

Further developments that have been awarded the Nobel Prize were the discovery of G-proteins by Martin Rodbell and Alfred Gilman, and cAMP by Earl Wilbur Sutherland. The later worked closely with De Duve who was later awarded the Prize for discovery of lysosomes.

References

Assan R, Hautecouverture G, Guillemand S, Dauchy F, Protin P, Dérot M. Evolution de paramètres hormonaux (glucagon, cortisol, hormone somatotrope) et énergétiques (glucose, acides gras libres, glycérol) dans dix acidocétoses diabétiques graves traitées.*Pathol. Biol.* 1969; 17: 1095–1098.

Blundell TL. The conformation of glucagon. In: Lefebvre PJ ed. Glucagon I. Berlin, Heidelberg, New York and Tokyo: Springer-Verlag, 1983; 37–56.

Bromer WW, Sinn LG, Staub A, Behrens OK. The amino acid sequence of glucagon. *J. Am. Chem. Soc.* 1956; 78: 3858–3859.

Bromer WW. Chemical characteristics of glucagon. In: Lefebvre PJ ed. Glucagon I. Berlin, Heidelberg, New York and Tokyo: Springer-Verlag, 1983; 1–21.

Brown, J. C., and J. R. Dryburgh. 1971. A gastric inhibitory peptide I. The amino acid composition and the tryptic peptides. *Can. J. Biochem.* 49:255-261.

Creutzfeldt W. The [pre-] history of the incretin concept. *Regulatory Peptides* 128 (2005) 87–91.

Deacon C., Andreas H. Johnsen, and Jens J. Holst. Degredation of glucagon like peptide-1 by human plasma in vitro yields an N-terminally truncated peptide that is a major endogenous metabolite in vivo. *JCEM* 1995;80:3,952-957.

Drucker DJ, Nauck MA (2006) The incretin system: glucagon-like peptide-1 receptor agonists and dipeptidyl peptidase-4 inhibitors in type2 diabetes. *Lancet* 368:1696–1705

Drucker DJ. Glucagon-like peptides. *Diabetes.* 1998;47(2):159-69.

Gregory. R.A. The Gastrointestinal hormone: an historical review. From http://www.pancreapedia.org/sites/www.pancreapedia.org/files/Gregory_Gastrointestinal_Hormones.pdf

Jelinek LJ, Lok S, Rosenberg GB, Smith RA, Grant FJ, Biggs S, Bensch PA, Kuijper JL, Sheppard PO, Sprecher CA, et al. Expression cloning and signaling properties of the rat glucagon receptor. *Science.* 1993 Mar 12;259(5101):1614-6.

John Eng et al. Isolation and Characterizationof Exendin-4, an Exendin-3 Analogue, from Heloderma suspecturn Venom. Further Evidence For An Exendinreceptor On Dispersed Acini From Guinea Pig Pancreas. *Journal of Biologicaclhemistr.*Vol. 267;11,7402-7406,1992.

Jornvall, H., A. Carlstrom, L. Pettersson, B. Jacobsson, M. Persson, and V. Mutt. 1981. Structural homologies between prealbumin, gastrointestinal prohormones and other proteins. *Nature* 291:261-263.

Kreymann B, Williams G, Ghatei MA, Bloom SR (1987) Glucagon-like peptide-1 7–36: a physiological incretin in man. *Lancet* 2:1300–1304.

Lee Y, Wang M-Y, Du XQ, Charron MJ, Unger RH. Glucagon receptor knockout prevents insulin-deficient type 1 diabetes in mice. *Diabetes* 2011; 60: 391–397.

Lefebvre P. J.. Early milestones in glucagon research. Diabetes, Obesity and Metabolism 13 (Suppl. 1): 1–4, 2011.

Lefebvre PJ. Le Glucagon, seconde hormone pancreatique. Paris: Maloine, 1967.

Liljestrand, G. (1952). The nobel prize in physiology and medicine. In Nobel the man and his prizes, pp. 135-316. Stockholm: the nobel foundation, sohlmans Forlag.

Loew ER, Gray JS, Ivy AC. Is a duodenal hormone involved in carbohydrate metabolism? *Am. J. Physiol.* 1940;129:659–63.

Loew ER, Gray JS, Ivy AC. The effect of acid stimulation of the duodenum upon experimental hyperglycemia and utilization of glucose. *Am. J. Physiol.* 1940;128:298–308.

Loew ER, Gray JS, Ivy AC. The effect of duodenal instillation of hydrochloric acid upon the fasting blood sugar of dogs. *Am. J. Physiol.* 1939;126: 270 – 6.

Mudaliar S and Henry. R. R. The incretin hormones: from scientific discovery to practical therapeutics. *Diabetologia* (2012) 55:1865–1868.

Murlin JR, Clough HD, Gibbs CBF, Stokes AM. Aqueous extracts of the pancreas. Influence on the carbohydrate metabolism of depancreatized animals. *J. Biol. Chem.* 1923; 56: 253–296.

MUTT, V., and S. I. SAID. 1974. Structure of the porcine vasoactive intestinal octacosa-peptide. The amino acid sequence. Use of Kallikrein in its determination. *Europ. J. Biochem.* 42:581-589.

Mutt, V., J. E. Jorpes, and S. Magnusson. 1970. Structure of porcine secretin. *Europ. J. Biochem.* 15:513-519.

Nauck M, Stöckmann F, Ebert R, Creutzfeldt W (1986) Reduced incretin effect in type 2 (non-insulin-dependent) diabetes. *Diabetologia* 29:46–52.

Nauck MA, Heimesaat MM, Orskov C, Holst JJ, Ebert R, Creutzfeldt W (1993) Preserved incretin activity of glucagon-like peptide 1 [7– 36 amide] but not of synthetic human gastric inhibitory polypeptide in patients with type-2 diabetes mellitus. *J. Clin. Invest.* 91:301–307.

Nauck MA, Vardarli I, Deacon CF, Holst JJ, Meier JJ (2011) Secretion of glucagon-like peptide-1 (GLP-1) in type 2 diabetes: what is up, what is down? *Diabetologia* 54:10–18.

Ringertz Nils, 1997. From *Nobel Lectures, Physiology or Medicine 1991-1995*.

Rouillé Y, Westermark G, Martin SK, Steiner DF. Proglucagon is processed to glucagon by prohormone convertase PC2 in alpha TC1-6 cells. *Proc. Natl. Acad. Sci. USA*. 1994 Apr 12;91(8):3242-6.

Spiess, J., J. Rivier, M. Thorner, and M. Vale. 1982. Sequence analysis of a growth hormone releasing factor from a human pancreatic islet tumor. *Biochemistry* 21:6037-6040.

Tricoli J V, Graeme I Bell, and Thomas B Shows. The Human Glucagon Gene Is Located on Chromosome 2. Diabetes February 1984 vol. 33 no. 2 200-202.

Unger RH, Aguilar-Parada E, Muller WA, Eisentraut AM. Studies of pancreatic alpha-cell function in normal and diabetic subjects. *J. Clin. Invest.* 1970;49: 837–848.

Unger RH, Leilo Orci. The essential role of glucagon in the pathogenesis of diabetes mellitus. *Lancet Jan.* 1975; 14-16.

White J. W, and Sauders G F. Structure of the human glucagon gene. *Nucleic Acids Res.* 1986 June 25; 14(12): 4719–4730.

Wunsch E, Weinges KF. The synthesis of glucagon. Properties of synthetic glucagon. In: Lefebvre PJ, Unger RH eds. Glucagon: Molecular Physiology, Clinical and Therapeutic Implications. Oxford: Pergamon Press, 1972; 31–46.

www.nobelprize.org/nobel_prizes/medicine/laureates/1994/presentation-speech.html

Zunz E, La Barre J. Contributions à l'étude des variations physiologiques de la sécrétion interne du pancréas: relations entre les sécretions externe et interne du pancréas. *Arch. Int. Physiol. Biochim.* 1929;31:20–44.

In: Glucagon
Editor: Ahmed Al-Jebawi

ISBN: 978-1-62618-147-2

Chapter 2

Glucagon Structure, Secretion and Function

Ahmed F. Al-Jebawi*
Indiana University School of Medicine,
Indianapolis, Indiana, US
Physician Training Program
at Dubai Diabetes Center, Dubai, UAE

Abstract

Glucagon has long been known as a critical hormone for glucose homeostasis. The recent finding of inappropriately elevated levels of glucagon in the circulation of diabetic patients has made a new revolution in the treatment of this disease. Glucagon is a 29-amino acid polypeptide produced by α-cells of the islets of Langerhans. Specific proteolytic enzymes are responsible for processing proglucagon precursors to glucagon and GLP-1 peptides. Glucagon effect on the carbohydrates and fat metabolism is quite complex. It stimulates glycogenolysis, gluconeogenesis, and ketogenesis by the liver. It also increases the rate of lipolysis in adipose tissue and glycogenolysis in muscle.

* Correspondence to: Ahmed F. Al-Jebawi, MD FACP FACE, E-mail: aljebawi@hotmail.com.

Introduction

The existence of a pancreatic chemical substance that potently stimulates blood glucose elevation was described just after the isolation of insulin. The 29-amino acid peptide glucagon was purified and sequenced in the mid-1950s and its role in normal and diabetic physiology was studied over the 5 decades since [1]. Despite its full purification and crystallization, glucagon clinical use was limited to the treatment of hypoglycemia for many years. In 1970, Muller et al. [2] described high glucagon levels in patients with type 2 diabetes. However, such finding did not receive much attention to develop therapeutic strategies for the treatment of type 2 diabetes. On the other hand, all medical treatment was focused on improving insulin secretion and action. It is now evident that the pathogenesis of diabetes is not limited to β-cell dysfunction and insulin resistance, but it also includes excessive glucagon production. Clinical studies show that high fasting glucagon levels and lack of suppression of postprandial glucagon secretion significantly contribute to the occurrence of hyperglycemia in type 2 diabetes [3]. Normally, hyperglycemia and hyperinsulinemia suppress glucagon secretion. However, in subjects with type 2 diabetes, such suppression is greatly impaired [4]. In vitro studies had shown that elevated plasma glucose levels in diabetic mice is accompanied by increased pancreatic α-cell number [5]. It is probable that in type 2 diabetes α-cells develop resistance to the inhibitory effects of insulin on glucagon secretion.

Physiological Role of Glucagon

Glucagon plays a primary role as a counterregulatory hormone. Through its coordinated interplay with the other glucose lowering and glucose raising factors, normal glucose homeostasis is maintained. Glucagon elevates plasma glucose levels by increasing both glycogenolysis and gluconeogenesis. In contrast, insulin suppresses hepatic glucose output while enhancing hepatic glucose uptake and glycogenesis, indicating that a balance between these two hormones determines hepatic glucose metabolism and thus systemic glycemic homeostasis. In addition to its effects on the liver, glucagon was shown to enhance the lipid breakdown in the adipose tissue while, in contrast, glucagon has little effect on skeletal muscle glucose metabolism [6]. Glucagon is a 29-amino acid peptide hormone encoded within a proglucagon precursor that also contains amino acid sequences for glucagon-like peptide (GLP)-1, GLP-2, oxyntomodulin, and glicentin (Figure 1). Proglucagon is expressed in the α-cells of the endocrine pancreas and in the enteroendocrine L-cells of the intestine. Due to the remarkable tissue-specific posttranslational expression of the proglucagon, the α-cells of the pancreas produce glucagon while L-cells of the intestine produce GLP-1. A post-translational cleavage by prohormone convertases (PC) is responsible for the maturation of the preproglucagon hormone that generates all these peptides. It is quite interesting that such processing of the identical proglucagon in the intestine and pancreas gives rise to peptides with opposing functions. In the α-cell, glucagon is stored in dense core secretory granules. Glucagon is released from the α-cell by Ca^{2+}-dependent exocytosis of secretory granules [7]. Glucagon secretion greatly depends on glucose levels. In wild-type mouse islets, glucose produced a concentration-dependent reduction of glucagon release [8]. It was clearly shown that hypoglycemia is a potent stimulus for glucagon release and hyperglycemia suppresses glucagon levels in pancreatic venous blood in dogs [9]. Hepatic clearance of glucagon is 20–30% of the portal content, with the kidney contributing to the major portion of peripheral glucagon removal [10].

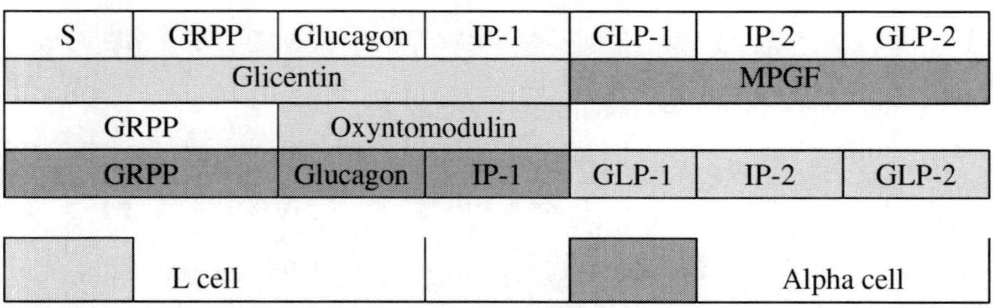

Figure 1. Schematic representation of tissue-specific post-translational processing of proglucagon in the intestinal L cell and the pancreatic α cell. Tissue-specific posttranslational processing of proglucagon in the pancreas gives glicentin-related polypeptide (GRPP), glucagon, intervening peptide-1 (IP-1), and major proglucagon fragment (MPGF), whereas glicentin, oxyntomodulin (OXM), intervening peptide (IP-2), glucagon-like peptide (GLP)-1, and GLP-2 are liberated from the brain and the intestine. Modified from Pharmacol Rev. Author manuscript; available in PMC 2009 June 15. Published in final edited form as: Pharmacol Rev. 2008 December; 60(4): 470–512. *Pharmacol Rev*. Kim and Egan Page 72 *Pharmacol Rev*. U.S. Government work not protected by U.S. copyright 604/3428101.

Glucagon and Hepatic Glucose Production

Glucagon plays a very critical role in regulating both the hepatic glucose uptake and hepatic glucose production. Net hepatic glucose uptake was markedly inhibited by concentrations of glucagon normally seen in the basal state. The increase in glucagon level among patients suffering from diabetes and inflammatory conditions further diminishes net hepatic glucose uptake [11]. The glucagon induced hepatic glucose production occurs through activating both the glycogenolytic and gluconeogenic pathways. Glucagon's acute effect on glucose levels is attributable to its glycogenolytic action. Hepatic glycogenolysis peaks at 15 min and falls back to baseline by 3 hours [12]

Glucagon's Role in Hypoglycemic Counterregulation

Glucagon provides the body's primary defense against hypoglycemia. The ability of glucagon to stimulate hepatic glucose output was found to increase by almost threefold when hypoglycemia with its associated increase in plasma catecholamines occurs [13].

Hypoglycemia is associated with a decrease in insulin signaling and a glucagon mediated increase in the activation of AMP-activated protein kinase (AMPK). These changes were in turn associated with an augmentation of glucagon's stimulatory effect on glycogen phosphorylase activity and its inhibitory effects on glycogen synthase activity.

The disruption in insulin signaling thereby allows glucagon's effect on cAMP-activated protein kinase (PKA) and AMPK to drive glycogen breakdown in a life-saving manner. PKA has long been established as the major intracellular mediator of glucagon's glucoregulatory effects within the hepatocyte, and prominently coordinates the phosphorylation of glycogen synthase and glycogen phosphorylase, resulting in net glycogen breakdown [13]

Glucagon and the Cardiovascular System

Glucagon produces robust and rapid metabolic effects in the myocardium [14]. Although it activates adenyl cyclase, such activation does not occur through the catecholamine beta-adrenoceptor stimulation [15, 16]. Nevertheless, adenyl cyclase activation, whether mediated via glucagon or catecholamines, generates cyclic adenosine monophosphate (cAMP) resulting in significant changes on the control of fuel metabolism. Both positive inotropic and chronotropic effects were described as clinical effects of glucagon injections [17].

Rather than opposing the effects of insulin, glucagon mimics rapid metabolic actions of insulin on the myocardium. It caused dose-dependent and marked stimulation of glycolysis in addition to increasing glucose oxidation and suppressing palmitate oxidation [17].

Both glucagon and insulin stimulate glycolysis in heart by activating the same signal responsible for mediating the glycolytic pathway [18]. The research in this field had indicated that physiological levels of glucagon produce insulin-like increases in cardiac glucose utilization [14]. Due to its positive inotropic and chronotropic effects and demonstrated efficacy, glucagon has been used in the treatment of acute beta-blocker overdose [19].

Glucagon and the Kidney

Like its action on the myocardium, glucagon stimulates adenylate cyclase and cAMP production in nephrons and in cell-free preparations of human renal medullas [20]. Glucagon regulates the rate of kidney filtration, urea excretion, and water reabsorption by the kidney via direct and indirect mechanisms [21]. On the other hand, long-term hyperglucagonaemia in mice can result in mesangial cell proliferation and hypertension [22].

Glucagon and the Endocrine Pancreas

Glucagon has been shown to regulate cAMP production in β-cells and induce insulin secretion [23]. However, glucagon-mediated cAMP production in β-cells is less potent than that induced by the incretin hormones GLP-1 and gastric inhibitory polypeptide (GIP) [24]. It was shown that α-cells contain glucagon receptors and that exposure of α-cells to glucagon leads to a concentration-dependent increase in intracellular cAMP and stimulation of exocytosis. [110].

Glucagon Action in the Brain

There is strong evidence that glucose sensing cells are present in the central nervous system. Glucose sensitive neurons are thought to be responsible for glucagon and insulin secretion through the activation of the sympathetic and parasympathetic branches of the autonomic nervous system. As the brain depends mostly on glucose as a source of metabolic energy, the major goal of maintaining euglycemia is to ensure a sufficient flux of this nutrient to the brain.

Following peripheral treatment, Glucagon was found to have a glycogenolytic action in rainbow trout brain through the same mechanism of action on liver cells. Despite the fact that the amount of brain glycogen is low, its levels decreased by up to 88% after glucagon was peripherally injected. Further evaluation showed that peripheral injection of glucagon decreased brain glycogen synthetase activity by up to 42% and increased brain glycogen phosphorylase activity by up to 30%. In addition, lactate increased by up to 60% and acetoacetate decreased by up to 67%. All of these changes occurred in a time- and dose-dependent fashion [25].

Drucker et al. reported that the glucagon gene is expressed in the brainstem and hypothalamus and that a glucagon mRNA transcript identical to that produced in pancreas and intestine gives rise to proglucagon-related peptides in the brain [26]. Glucagon binds to brain membranes and mouse astrocytes and stimulates adenylate cyclase and cAMP production, respectively [27, 28]. Administration of glucagon was shown to reduce feeding [29, 30]. In another observation, glucagon was shown to reduce the meal size in humans after it was intravenously infused [31].

The subjects who received intravenous glucagon infusion reported feeling equally satiated as during control meals and reported no side effects, and the dose administered produced systemic plasma glucagon concentrations that approximated the range produced by normal meals [32]. To further prove this association, Langhans et al. demonstrated that reducing glucagon signal is associated with increased food intake. This conclusion was reached after they found out that the intraperitoneal administration of a highly specific polyclonal glucagon antibody was followed by a significantly increased meal size [33]. Since total subdiaphragmatic vagotomy, selective hepatic vagotomy, or lesion of the vagal afferent terminal fields all prevent glucagon's feeding-inhibitory effect, glucagon is thought to send signals to reduce meal size through sensory axons of the vagus nerve [34, 35]. The anorectic effects of glucagon may also involve suppression of ghrelin secretion. Glucagon, after intramuscular administration, significantly decreased circulating ghrelin in healthy subjects. However, glucagon failed to yield any significant effect in patients with a hypothalamic-pituitary lesion indicating that such actions require an intact hypothalamic-pituitary axis [36]. Patients suffering from glucagonoma usually have extremely elevated glucagon levels and severe anorexia [37, 38].

Glucagon's Role in Exercise

Physical exercise elicits a complex pattern of adaptive neuroendocrine and metabolic responses aimed at maintaining glucose homeostasis in the face of increasing energy substrate needs. If adequately activated, these responses, including increased secretion of glucagon, catecholamines, cortisol and growth hormone can effectively prevent the onset of hypoglycemia. It was shown that the exercise induced increment in glucagon controls 68% of the rise in hepatic glucose production during muscular work when the normal fall in insulin is present [39].This large effect of glucagon on the glucose release from the liver contributes markedly to the extra glucose produced during exercise. Glucagon deficiency in patients suffering from type 1 diabetes may explain their high risk of developing hypoglycemia during exercise [40]. Therefore, the role of glucagon in protecting the muscle and CNS from hypoglycemia during exercise is very critical.

Glucagon's Effect during an Acute Stress

Glucagon is markedly elevated during an acute stress [41]. Infection leads to profound alterations in whole-body metabolism, which is characterized by marked acceleration of glucose metabolism. Hyperglycemia was shown to be one of the characteristic effects of infection [42]. Hyperglycemia occurs due to a combined increase in glucose production and a failure of peripheral glucose clearance to increase in proportion to the increase in glucose production [43]. The rise in glucagon observed during endotoxemia increases hepatic glucose metabolism [44]. Following the acute administration of endotoxin, glycogenolysis is the primary source of the glucose [45], but during chronic infections, if glycogen stores are depleted, an increase in gluconeogenesis contributes to the enhancement of glucose production [46]. The acute increases in glucagon reduce liver glucose uptake by stimulating hepatic glucose production and inhibiting glycogen synthesis [48]. It is likely that glucagon enhances the hepatic glucose production in acute stressful conditions where impaired hepatic metabolism is a major problem. As a result, the risk of hypoglycemia is significantly reduced.

Table 1. Glucagon Effects

Effects of Glucagon on the cardiovascular system:

- Exerts positive cardiac inotropic effect
- Enhances glucose oxidation.
- Stimulates glycolysis.

Effects of Glucagon on the central nervous system:

- Induces glucose production
- Promotes satiety

Effects of Glucagon on the kidney:

- Increases rate of glomerular filtration and water re-absorption.

Effects of Glucagon on adipose tissue:

- May increase the rate of lipolysis.

Effects of Glucagon on the islet cells:

- Stimulates insulin secretion by beta cells.
- Autoregulates alpha cells.

Effects of Glucagon on the liver:

- Promotes glycogenolysis.
- Stimulates gluconeogenesis.
- Inhibits glycogenesis.
- Increases lipid oxidation.

Effects of Glucagon on the intestines:

- Inhibits gut motility.

Physiological Regulation of Glucagon

Physiological regulation of glucagon secretion is mediated through paracrine, hormonal and neuronal mechanisms. The term “paracrine" refers to such local interactions between different cells within the same islet. The paracrine effect was discovered after the observation that glucose stimulates rather than inhibits glucagon release in rat α-cells that were isolated from the other pancreatic cells [49]. After the introduction of insulin, somatostatin or certain other agents that are usually produced by the neighboring β- and δ-cells, such glucose stimulated glucagon secretion is counteracted [50]. The paracrine effect is greatly facilitated by a unique histological arrangement of glucagon-secreting α-cells, insulin-producing β-cells, somatostatin-releasing δ-cells and the other endocrine cells of the islet of Langerhans. Such cytoarchitecture of human pancreatic islets has important implications for islet cell function. Unlike the architecture of rodent islets, human α–cells are arranged in close proximity with β -cells allowing them to have very active paracrine interactions [51]. The vascular organization within the islets plays a major role to sustain such interactions by providing an important intercellular communication. A leading hypothesis to explain the interaction between α and β-cells is based on an earlier observation that the blood flow within the pancreatic islet is unidirectional. β-cells in the islet core receive the arterial blood before the peripherally located α-cells. According to this hypothesis, upstream β-cell products influence the function of downstream α-cells. Despite the fact that an extensive research was done to know the mechanism of glucagon regulation, this subject is still highly debatable.

Paracrine and Hormonal Regulation of Glucagon Secretion

Many factors were shown to play a certain role in glucagon production. Nevertheless, only the major physiological regulators will be discussed in this review. The most important among those are insulin, Zinc, GABA, GLP-1, somatostatin, ghrelin, glutamate, amylin and glucagon itself.

A. *Insulin*: Insulin role in glucagon regulation was demonstrated in many observations. Banarer S et al. showed that intraislet hyperinsulinemia prevents the glucagon response to hypoglycemia despite an intact autonomic response [52]. A similar effect was observed with the exogenous addition of insulin [53]. Many other studies support this observation by showing that insulin suppresses glucagon secretion when glucose levels are high or normal. However, when insulin level becomes low as a result of hypoglycemia, glucagon secretion is stimulated [54]. *Greenbaum* et al. showed that in drug-induced insulin deficiency in dogs, glucose failed to suppress glucagon secretion in the absence of exogenous insulin [55].

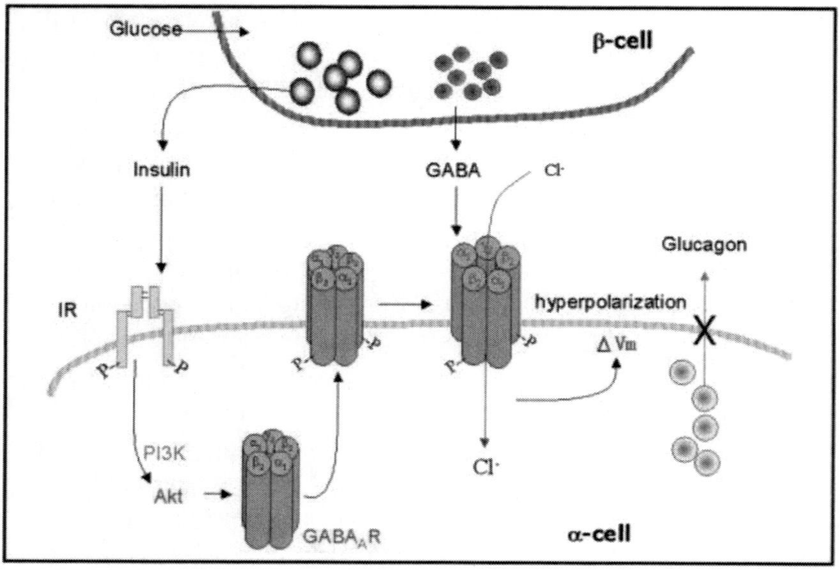

Figure 2. A model showing the Akt-GABA$_A$R pathway in mediating glucose-induced suppression of glucagon release in α-cells. When glucose levels increase, insulin released from β cells activates the IR present on α-cells. Activation of Akt leads to phosphorylation of β subunits of GABA$_A$R that causes rapid translocation of the receptor to the plasma membrane. As GABA is constantly released from β cells, increased GABA$_A$R numbers at the cell surface increases the efficacy of the receptor-mediated inhibitory currents (Cl^-) and membrane hyperpolarization occurs. In turn, membrane hyperpolarization shuts down voltage-dependent Ca^{2+} channels and as a result inhibits α cell exocytosis and glucagon release. The Ca^{2+}-dependent exocytosis machinery is not shown. From Elaine Xu, Mohan Kumar, Yi Zhang, William, Toshiyuki Obata, Nina Zhang, Shiying Liu, Anna Wendt, Shaoping Deng, Yousuke Ebina, Michael B. Wheeler,1 Matthias Braun, and Qinghua Wang. Intra-islet insulin suppresses glucagon release via GABA-GABAA receptor system. CELL METABOLISM 3, 47–58, JANUARY 2006 (Adapted with permission).

These results support the concept that intra-islet insulin action is essential for suppression of glucagon in response to hyperglycemia. Like their fundamental role in β-cells, ATP-dependent K^+ (K_{ATP}) channels are very important to maintain the function of α-cells. By activating K_{ATP}-channel, insulin induces membrane hyperpolarization of the α-cells resulting in a transient inhibition of electrical activity and a consequent suppression of glucagon secretion [56]. Insulin effect on membrane hyperpolarization and suppression of glucagon production was also shown to be mediated by activating GABA$_A$ receptors in the α-cells [57]. It is quite interesting that action potential normally occurs at low glucose levels [58].

As soon as the extracellular glucose levels rise, the intracellular ATP/ADP ratio increases leading to the blockade of K_{ATP} channels [59]. Although the paracrine effect was considered as the primary mechanism responsible for glucagon regulation, Mac Donald et al. have recently shown that there is a minimal direct inhibitory effect of glucose on the α-cells before β-cell activation and insulin secretion [60]. It was implicated that α-cells respond to glucose levels through a special glucose-sensing pathway similar to that found in insulin secreting β-cells, involving closure of ATP dependent K channels in the presence of glucose. Nevertheless, the direct inhibitory action of glucose on α-cell secretion seems to be of little physiological significance compared with the large paracrine and hormonal effects. While α-cells are inhibited by physiological increases in glucose levels, supraphysiological glucose levels were recently reported to paradoxically stimulate the glucagon secretion [61].

This phenomenon should be clinically significant because diabetes is associated with both hyperglycemia and hyperglucagonemia. On the other hand, diabetic patients frequently suffer from impaired glucagon release during hypoglycemia [62]. The paradoxical increase in glucagon levels among diabetic patients might result from reduced channel's ATP-sensitivity [63, 64] or increased K_{ATP}-channel activity associated with an age-dependent deterioration of mitochondrial function and impaired ATP production [65].

***B. Zinc*:** Zinc appears to play another significant role in the regulation of glucagon secretion. As Zinc is co-secreted along with insulin from β-cells, it was thought that it could be the active agent responsible for glucagon inhibition during periods of β-cell stimulation. By opening α-cell K_{ATP} channels, Zinc was shown to cause tonic suppression of glucagon secretion [66]. However, during states of hypoglycemia, β-cells virtually cease releasing zinc-insulin. This results in cessation of zinc effects on α-cell K_{ATP} channels and closure of the channels, which in turn causes depolarization, voltage-dependent calcium channel opening, calcium entry, and glucagon exocytosis (Figure 3, 4) [66].

C. GABA: γ-Aminobutyric acid (GABA) is a neuronal inhibitory transmitter that was proposed to function as a paracrine signaling molecule in islets of Langerhans. It was shown that rat β-cells release GABA by Ca^{2+}-dependent exocytosis of synaptic-like microvesicles. When GABA is released, it can diffuse over sufficient distances within the islet interstitium to activate $GABA_A$ receptors in neighboring glucagon-secreting α-cells. Upon $GABA_A$ receptor activation, the electrical activity gets suppressed with a consequent reduction in exocytosis of the glucagon-containing secretory granules in the α-cell. [67]

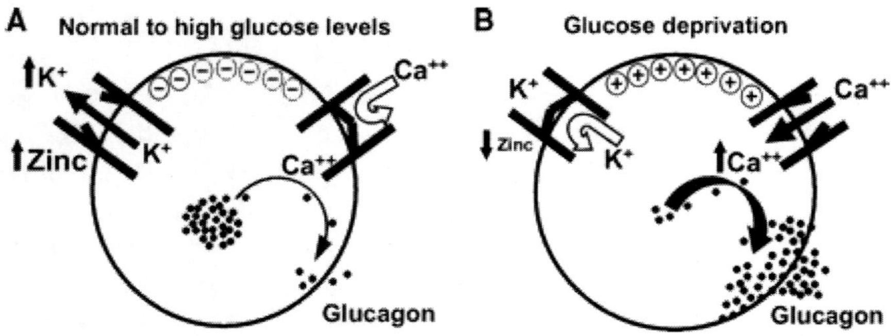

Figure 3. Regulation of glucagon secretion by zinc. *A*: α-Cell electrical and hormonal status in states of physiologic and elevated glucose conditions. β-Cells release zinc and insulin hexamers into the intraislet periportal circulation. Zinc dissociates from insulin and reaches downstream α-cells, where it binds to and opens the K_{ATP} channels. K^+ ions leave the cell and hyperpolarize the α-cell, thus preventing voltage-dependent calcium channels from opening. Glucagon granules are not mobilized and remain stored inside the cell. *B*: When blood glucose levels decrease in response to hypoglycemic levels, β-cell insulin and zinc secretion decrease as well. K_{ATP} channels on α-cells close, K^+ remains in the cell, and the α-cell depolarizes, which induces calcium channels to open and calcium enters the cell. Intracellular calcium rises, inducing glucagon exocytotic granules to migrate to the plasma membrane, where they fuse and release glucagon into the portal venous system. From: Michela Slucca, Jamie S. Harmon, Elizabeth A. Oseid, Joseph Bryan andR. Paul Robertson. ATP-Sensitive K^+ Channel Mediates the Zinc Switch-Off Signal for Glucagon Response During Glucose Deprivation. DIABETES, JANUARY 2010. VOL. 59: 128-134 (with permission).

D. *GLP-1*: GLP-1 is a 30-amino acid peptide produced by L cells in the lower intestine by alternative processing of the gene encoding proglucagon [68]. Physiologically, L-cells, which are most common in the distal parts of the intestines [69], are stimulated to secrete all their secretory products (GLP-1, GLP-2, gut-derived glucagon [also called glicentin], peptide tyrosine, oxyntomodulin) in response to nutrient ingestion [70, 71].

By definition, all gut hormones known to stimulate insulin secretion in response to food ingestion are called incretins. Incretins play a pivotal role in regulating glucose homeostasis. GLP-1 is considered as an important incretin hormone that is secreted into the circulation by L cells of the small intestine. The incretin effect was first described in the 1960s during an experiment where similar elevations in plasma glucose resulted in a greater insulin secretory response when glucose was administered orally rather than intravenously [72, 73].

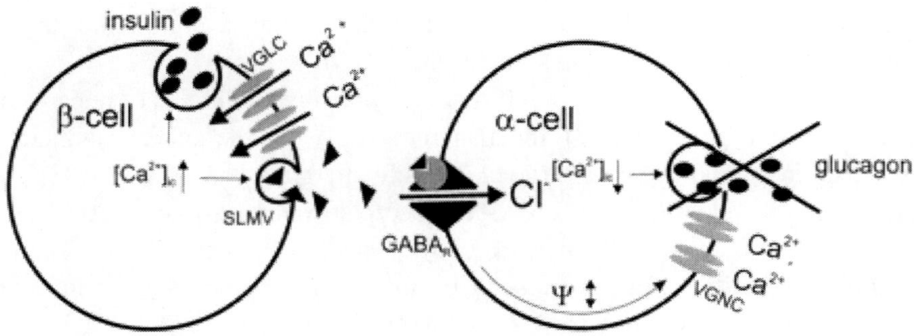

Figure 4. Model for GABAergic regulation of glucagon secretion. See text for details. $[Ca^{2+}]_i$, cytoplasmic free Ca^{2+} concentration; $GABA_R$, $GABA_A$ receptor Cl^- channels; VGLC, voltage-gated L-type Ca^{2+} channels; VGNC, voltage-gated N-type Ca^{2+} channels; ψ, membrane potential; ↕, an increase or decrease in ion concentration or membrane potential (depolarization/hyperpolarization). From: Anna Wendt, Bryndis Birnir , Karsten Buschard, Jesper Gromada, Albert Salehi, Sabine Sewing, Patrik Rorsman and Matthias Braun. Glucose Inhibition of Glucagon Secretion From Rat α-Cells Is Mediated by GABA Released From Neighboring β-Cells. DIABETES, VOL. 53, APRIL 2004 (with permission).

The main meal components that act as potent stimulants of GLP-1 secretion are glucose [74] and triacylglycerol, but fructose and some proteins are also effective [75].

As a matter of fact, GLP-1 secretion can occur in advance of the direct contact between food and the L cells, suggesting that a neural and/ or hormonal communication pathway triggers the release of GLP-1 to prepare the body in advance of the absorption of carbohydrates from the meal [76]. The inhibitory effect of GLP-1 on glucagon release from α-cells was proved [77]. Such effect can be mediated through a direct hormonal influence [78] or by the paracrine effect of the neighboring α- or β-cells [79]. The inhibitory effect of GLP-1 on glucagon secretion is not only caused by increasing the local concentration of insulin, since such inhibition was shown to occur even when insulin was unstimulated [80, 81]. While GLP-1 causes glucagon secretion inhibition during modest hyperglycemia, it does not inhibit the increase in glucagon levels in response to hypoglycemia [82]. This selective effect of GLP-1 plays a major role in protecting diabetic patients from having hypoglycemia while they are receiving a GLP-1 agonist as a monotherapy or in combination with metformin or a thiazolidinedione.

E. *Somatostatin*: In addition to its storage in pancreatic δ cells which comprise 5–10% of the islet endocrine cells [83], somatostatin is synthesized and secreted by neuroendocrine cells in the central nervous system and the gastrointestinal system. In fact, gastrointestinal system is considered as the major contributor to circulating somatostatin [84, 85].

This peptide hormone was shown to potently inhibit the release of glucagon from the α-cells of the pancreas [86]. On the other hand, anti-somatostatin antibodies were found to activate glucagon release [87]. In addition to its identification on α-cells, somatostatin receptors have been identified on β-cells [88]. It was shown that exogenously administered somatostatin inhibits glucose-induced insulin secretion and arginine-induced glucagon secretion both in vitro and in vivo [89].

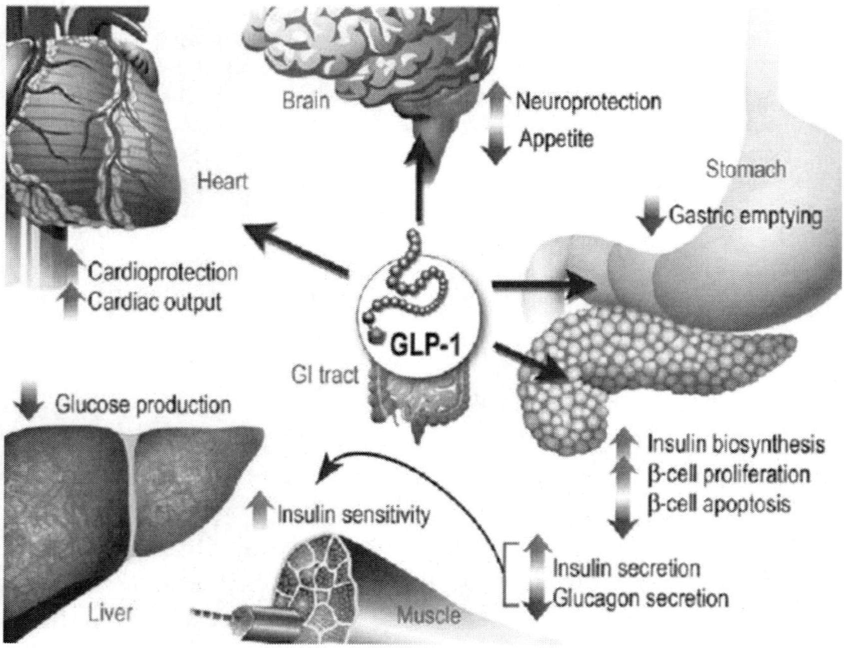

Figure 5. Actions of glucagon-like peptide 1 (GLP-1) in peripheral tissues. The GLP-1 acts directly on the endocrine pancreas, heart, stomach, and brain, whereas actions on liver and muscle are indirect. GI = gastrointestinal. From: Drucker DJ. The biology of incretin hormones. *Cell Metab*. 2006;3(3): 153-165 (with permission).

Hauge-Evans et al. showed that arginine-induced release of insulin and glucagon is markedly stimulated in vivo when somatostatin is absent. Moreover, in islets isolated from somatostatin depleted δ-cells of mice, the normal inhibition of glucagon release by glucose was eliminated, while the stimulation of insulin release by the sugar was enhanced [90]. This dual effect of somatostatin on α- and β-cells may represent an important role in regulating responses to receptor-operated stimuli. Electrophysiological studies have shown that somatostatin activates K^+ channels in α-cells, inducing membrane hyperpolarization and suppressing electrical activity, which affects Ca^{+2}-dependent exocytosis [91, 92].

The inhibitory effect of somatostatin on glucagon secretion reveals the importance of δ-cells in the normal function of the endocrine pancreas and suggests that δ-cell dysfunction in diabetes may have important consequences for hormone secretion from islet α- and β-cells [93].

***F. Glutamate*:** Amino acids such as arginine, alanine and glutamine are potent stimulators of glucagon secretion. $_{L}$-Glutamate was shown to work as an autocrine signaling molecule in α- cells providing positive feedback for glucagon secretion in human islets. Glutamate is secreted by α-cells and is responsible for activating ionotropic glutamate receptors (iGluRs) on α-cells themselves. This positive feedback for α-cell function amplifies glucagon secretion and represents a typical example for autocrine signaling where a modest decrease in blood glucose concentration can effectively induce glucagon release from pancreatic α-cells [94, 95]. What makes this process more facilitated is the expression of vesicular glutamate transporters in α-cells. With the availability of these transporters, glutamate uptake into secretory vesicles becomes much easier [96]. To prove the importance of glutamate in glucagon production, it was shown that blockade of iGluRs diminishes the glucagon response to a lowering in glucose concentration [94]. These results indicate that glutamate signaling provides a positive autocrine feedback loop. Despite the fact that recent studies had shown that glutamate, acting on a cell metabotropic glutamate receptor type 4 (mGluR4), inhibits glucagon secretion, such inhibition was found to be minimal or absent in humans (and monkeys and mice): in these latter species, the positive autocrine action on α- cell iGluRs seems to dominate and promote glucagon release [97].

G. *Ghrelin*: Ghrelin is a 28-amino acid peptide that is secreted from the gastrointestinal tract. Many researchers believe that ghrelin is also expressed in pancreatic islets [98-100]. Ghrelin is synthesized as part of a larger precursor protein, preproghrelin that produces acyl ghrelin and des-acyl ghrelin [101, 102]. Acylation by the enzyme ghrelin O-acyltransferase (GOAT) enables ghrelin to activate the ghrelin receptor with an increase in food intake [103, 104] and growth hormone (GH) secretion [105]. Ghrelin was shown to increase plasma glucagon secretion in a dose-dependent manner from mouse islets and α-cell lines. Regulation of blood glucose through this hormone was found to involve a direct stimulation of glucagon secretion from α-cells.

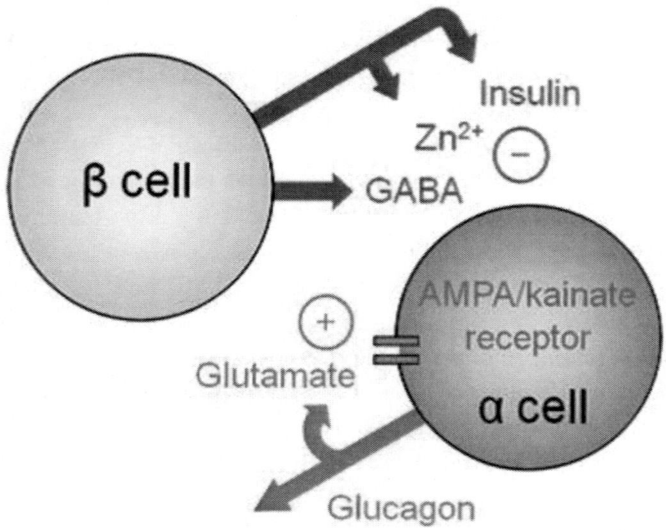

Figure 6. Proposed model for the regulation of glucagon secretion. Activation of α-cells depends on an initial stimulus as well as on positive feedback. When glucose levels fall, there is less suppression from b cell-derived GABA, Zn2+, or insulin. Positive feedback by glutamate strongly amplifies glucagon secretion. Once glucose levels increase, glucagon secretion is inhibited by insulin, Zn2+, GABA, or a combination of the three. Without glutamate feedback, α-cells are not fully activated and glucagon secretion is deficient. From: *Over Cabrera, M. Caroline Jacques-Silva, Stephan Speier, Shao-Nian Yang,Martin Ko¨ hler, Alberto Fachado, Elaine Vieira, Juleen R. Zierath, Richard Kibbey, Dora M. Berman, Norma S. Kenyon, Camillo Ricordi, Alejandro Caicedo and Per-Olof Berggren: Glutamate Is a Positive Autocrine Signal for Glucagon Release. Cell Metabolism 7, 545–554, June 2008 (with permission).*

Only few studies have reported ghrelin effects on insulin levels in humans. They have shown a decreased or no effect on fasting insulin and decreased intravenous glucose-stimulated insulin secretion. Nevertheless, the involvement of ghrelin in regulating glucose metabolism remains to be elucidated.

H. *Glucagon*: Recent experiments have revealed that the α-cell contains glucagon receptors and that exposure of α -cells to glucagon leads to a concentration-dependent increase in intracellular cAMP and stimulation of exocytosis [110].

Glucagon can paradoxically stimulate Ca^{+2}-dependent exocytosis in α-cells leading to more secretion of glucagon. The influence of glucagon on α-cell activity is another example of the autocrine regulation of these cells.

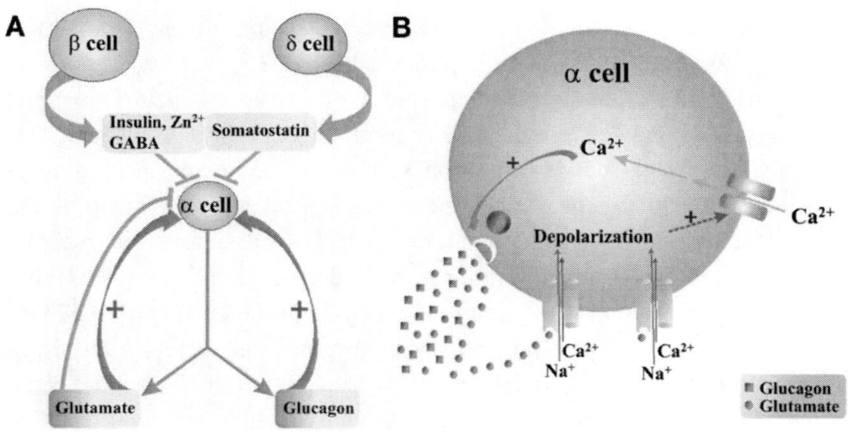

Figure 7. Paracrine and Autocrine Regulation of α Cell Secretion. (A) α-cell secretion is under negative paracrine control by β and δ cells. Glucagon released by α-cells provides a positive autocrine signal, while glutamate may exert either a negative or positive autocrine effect depending on the receptor subtype on which glutamate acts (see text). (B) In α-cells, activation of ionotropic glutamate receptors (iGluRs) of the AMPA and kainate subtypes by glutamate results in Na+ and Ca2+ influx, subsequently leading to membrane depolarization, opening of voltage-gated Ca2+ channels, and Ca2+-triggered secretory granule exocytosis. Note that for the AMPA and kainate receptor subtypes, Na+ permeability is considerably higher than Ca2+ permeability. From: Herbert Y. Gaisano and Yuk M. Leung: Pancreatic Islet a Cell Commands Itself: Secrete More Glucagon. Cell Metabolism 7, June 2008:474-475 (with permission).

J. ***Amylin*:** Amylin is a 37 amino acid peptide, which is stored and released together with insulin in response to nutrient stimuli [111]. It was recently shown that amylin reduces plasma glucagon levels in cats, a feature that has treatment potential in diabetic animals [112]. Earlier results from studies with the amylin agonist pramlintide in patients with type-1 diabetes suggest that amylin influences postprandial glucose levels by decreasing glucagon secretion [113].

Autonomic Regulation of Glucagon Secretion

The pancreatic islets are richly innervated by parasympathetic, sympathetic and sensory nerves. Glucose is by far the predominant metabolic fuel used by the brain [114]. The brain cannot synthesize glucose, nor can it store more than a few minutes' supply as glycogen. Therefore, the brain is dependent on a continuous supply of glucose from the circulation. Several different neurotransmitters are stored within the terminals of these nerves, including the classical neurotransmitters, acetylcholine and noradrenaline in addition to several other neuropeptides [115, 116]. It was shown that stimulation of the autonomic nerves and treatment with neurotransmitters affect islet hormone secretion of glucagon [117, 118]. It has also been established that vagal nerve activation stimulates the secretion of glucagon [119]. The sympathetic nervous system plays a major role in stimulating glucagon secretion, through activation of the α-cell adrenergic receptors. Sympathetic system activation induces the adrenal glands to release more epinephrine through the blood stream. High epinephrine levels combined with nervous secretion of norepinephrine stimulate glucagon secretion and inhibit insulin secretion [120]. Epinephrine was also shown to enhance hepatic glucose output and decrease glucose clearance by tissues such as muscle and fat. Hypothalamus and brainstem play a primary role in regulating the activation or inhibition of either the parasympathetic or sympathetic nervous system through special glucose-sensing cells [121-123]. The activity of the sympathetic and parasympathetic nervous system is also determined by the input of the glucose-sensing cells located at different anatomical sites including the mouth, the gut, the carotid body, the hepatoportal vein area, the brainstem and hypothalamus. During hypoglycemic stresses, glucagon secretion is stimulated considerably whereas insulin secretion is inhibited, providing a mechanism for the recovery of circulating glucose through stimulation of hepatic glucose output and inhibition of peripheral glucose uptake [124]. Several studies have shown that the islet nerves are activated during the hypoglycemic stress, thus providing a basis for neural contribution to the changes in islet hormone secretion [124]. Atropine and adrenergic blockade have also been shown to inhibit hypoglycemia-induced glucagon secretion, further proving the role of neurogenic regulation of glucagon [124]. The ventromedial hypothalamus (VMH) has been shown to play an important role in sensing decreases in blood glucose levels and initiating counterregulatory responses to hypoglycemia [126-130]. Eating stimulates an immediate, brief increase in glucagon secretion. It occurs within one minute of the onset of spontaneous meals [131]. This observation, coupled with the absence of prandial glucagon secretion in patients with transplanted pancreata, [132] imply that glucagon secretion can be elicited by cephalic stimuli. Although most meals, except pure carbohydrate meals, stimulate glucagon secretion, high-protein meals are most effective [133, 134]. The frequently seen impaired glucagon response to hypoglycemia among patients with type 1 diabetes could be caused by autonomic nervous system failure to regulate this process.

Defects in Glucagon Secretion in Type 2 Diabetes

The defects of glucagon secretion associated with diabetes in man are twofold: excessive amounts of glucagon being secreted at high glucose levels and low amounts at low glucose levels [135-137]. Indeed, glucagon secretion may be paradoxically increased during a glucose challenge [138]. In addition to insulin deficiency and insulin resistance, patients with type 2 diabetes have a blunted or absent α-cell response to hyperglycemia and their plasma glucagon remains inappropriately elevated [139]. Moreover, glucagon secretion is increased by normal α-cell stimuli, such as IV arginine or protein-rich meals, to a greater extent in type 2 DM than non-diabetic individuals [140,141]. Fasting and postprandial hyperglucagonemia in diabetic patients is considered as a main factor for increased hepatic glucose output both before and after meals [142-144]. The defect in alpha cell function that occurs in type 2 diabetes reflects impaired glucose sensing. It has also been suggested that the α-cell is resistant to insulin or other β-cell secretory products such as zinc or GABA. [145]. However, it is still quite difficult to distinguish between impaired glucose sensing and α-cell insulin resistance with type 2 diabetes. Recently, the ratio of alpha to beta cell areas in pancreas samples obtained at autopsy was measured from 50 type 2 diabetic subjects, whose beta cell mass had previously been found to be 36% lower than that of 52 non-diabetic subjects. The average ratio was significantly higher in type 2 diabetic subjects (0.72) than in non-diabetic subjects (0.42), with, however, a large overlap between the two groups. In contrast, the alpha cell mass was virtually identical in type 2 diabetic subjects and non-diabetic subjects. Based on these results, it was presumed that the higher proportion of alpha to beta cells in the islets of type 2 diabetic subjects is due to a decrease in beta cell number rather than an increase in alpha cell number. This imbalance may contribute to alterations in the normal inhibitory influence exerted by beta cells on alpha cells, and lead to the relative hyperglucagonaemia observed in type 2 diabetes [146]. This finding is consistent with the concept that the reason for glucagon overproduction is beta cell failure and insulin deficiency rather than alpha cell growth. [147]. Mice with a targeted disruption of the glucagon receptor gene were found to have significantly lower fasting and postprandial glucose levels relative to either their nontransgenic littermates or wild-type mice [148]. While patients with type 1 diabetes have a major problem with loss of glucagon stimulation during episodes of hypoglycemia [149], most studies done on type 2 patients revealed an impaired response of glucagon secretion to the stimulatory effect of hypoglycemia especially when type 2 diabetes is advanced [150-153]. As the glucose suppressive effect on glucagon production was also found to be impaired in subjects with IGT [154], the question whether hyperglucagonemia is a contributing factor for the etiology of type 2 diabetes becomes critical.

Conclusion

Glucagon is a key counterregulatory hormone produced by α-cells of the endocrine pancreas. It plays a major role in glucose homeostasis by increasing both glycogenolysis and gluconeogenesis. Physiological regulation of glucagon secretion is mediated through paracrine, hormonal and neuronal mechanisms. The most important among the paracrine regulators are insulin, Zinc, GABA, GLP-1, somatostatin, ghrelin, glutamate, amylin and glucagon itself. Islet hormone secretion of glucagon was shown to increase with stimulation of the autonomic nerves and treatment with certain neurotransmitters. Patients with type 2 diabetes were found to have inappropriately elevated glucagon levels. Furthermore, their glucagon response to hypoglycemia is impaired. Although it seems that the reason for glucagon overproduction is beta cell failure and insulin deficiency rather than alpha cell growth, further research is needed in this field.

References

[1] Staub A, Sinn L, Behrens OK 1955 Purification and crystallization of glucagon. *J. Biol. Chem.* 214:619–632.

[2] Muller WA, Faloona GR, Aguilar-Parada E, Unger RH 1970 Abnormal cell function in diabetes. Response to carbohydrate and protein ingestion. *N. Engl. J. Med* 283:109–115.

[3] Baron AD, Schaeffer L, Shragg P, Kolterman OG (1987) Role of hyperglucagonemiain maintenance of increased rates of hepatic glucose output in type IIdiabetics. *Diabetes* 36: 274–283.

[4] Gerich JE, Charles MA, Grodsky GM 1976 Regulation of pancreatic insulin and glucagon secretion. *Ann. Rev. Physiol.* 38:353–388.

[5] Zhuo Liu, Wook Kim, Zhike Chen, Yu-Kyong Shin, Olga D. Carlson, Jennifer L. Fiori, Li Xin, Joshua K.Napora, Ryan Short, Juliana O. Odetunde, Qizong Lao, Josephine M. Egan. Insulin and Glucagon

Regulate Pancreatic a-Cell Proliferation. *PLoS ONE* January 2011 | Volume 6, Issue 1, e16096.

[6] Christophe J. Glucagon and its receptor in various tissues. *Ann. N Y Acad. Sci.* 1996; 805: 31–42; discussion 42–33.

[7] Gromada, J. et al. (1997) Adrenaline stimulates glucagon secretion in pancreatic A-cells by increasing the Ca2+ current and the number of granules close to the L-type Ca2+ channels. *J. Gen. Physiol.* 110, 217–228.

[8] Jesper Gromada, Xiaosong Ma, Marianne Høy, Krister Bokvist, Albert Salehi,Per-Olof Berggren, and Patrik Rorsman. ATP-Sensitive K_ Channel–Dependent Regulation of Glucagon Release and Electrical Activity by Glucose in Wild-Type and SUR1_/_ Mouse _Cells. Diabetes, Vol. 53, Supplement 3, December 2004.

[9] Unger RH, Eisentraut AM, McCall MS, Madison LL 1962 Measurements of endogenous glucagon in plasma and the influence of blood glucose concentration upon its secretion. *J. Clin. Invest.* 41:682–689.

[10] Dobbins RL, Davis SN, Neal DW, Cobelli C, Jaspan J, Cherrington AD. Compartmental modeling of glucagon kinetics in the conscious dog. *Metabolism* 1995; 44: 452–459.

[11] Chen SS, Zhang Y, Santomango TS, Williams PE, Lacy DB, McGuinness OP. Glucagon chronically impairs hepatic and muscle glucose disposal. *Am. J. Physiol. Endocrinol. Metab.* 2007 Mar;292(3):E928-35.

[12] Ramnanan C. J., Edgerton D. S., Kraft G. and Cherrington A. D. Physiologic action of glucagon on liver glucose metabolism. *Diabetes, Obesity and Metabolism 13* (Suppl. 1): 118–125, 2011].

[13] Rivera N, Ramnanan CJ, An Z et al. Insulin-induced hypoglycemia increases hepatic sensitivity to glucagon in a canine model. *J. Clin. Invest.* 2010; 120: 4425–4435.

[14] Harney JA, Rodgers RL. Insulin-like stimulation of cardiac fuel metabolism by physiological levels of glucagon: involvement of PI3K but not cAMP. *Am. J. Physiol. Endocrinol. Metab.* 295: E155–E161, 2008.

[15] Pollack CV: Utility of glucagon in the emergency department. *J. Emerg. Med.* 1993;11:195-205.

[16] Sener EK, Gabe S, Henry JA: Response to glucagon in imipramine overdose. *Toxicol.-Clin. Toxicol.* 1995;33:51-53.

[17] Muhlhauser I, Koch J, Berger M: Pharmacokinetics and bioavailability of injected glucagon: differences between intramuscular, subcutaneous, and intravenous administration. *Diabetes Care* 1985;8:39-42.

[18] Lefebvre V, Mechin MC, Louckx MP, Rider MH, Hue L. Signaling pathway involved in the activation of heart 6-phosphofructo-2-kinase by insulin. *J. Biol. Chem.* 271: 22289–22292, 1996.

[19] Weinstein RS: Recognition and management of poisoning with beta-adrenergic blocking agents. *Ann. Emerg. Med.* 1984;13:1123-1131.

[20] Mulvehill JB, Hui YS, Barnes LD, Palumbo PJ, Dousa TP. Glucagonsensitive adenylate cyclase in human renal medulla. *J. Clin. Endocrinol. Metab*. 42: 380–384, 1976.

[21] Gustavson SM, Chu CA, Nishizawa M, Neal D, Farmer B, Yang Y, Donahue EP, Flakoll P, Cherrington AD. Effects of hyperglycemia, glucagon, and epinephrine on renal glucose release in the conscious dog. *Metabolism* 53: 933–941, 2004.

[22] Li XC, Liao TD, Zhuo JL. Long-term hyperglucagonaemia induces early metabolic and renal phenotypes of type 2 diabetes in mice. *Clin. Sci.* (Lond) 114: 591–601, 2008.

[23] Kieffer TJ, Heller RS, Unson CG, Weir GC, Habener JF. Distribution of glucagon receptors on hormone-specific endocrine cells of rat pancreatic islets. *Endocrinology* 137: 5119–5125, 1996.

[24] Moens K, Heimberg H, Flamez D, Huypens P, Quartier E, Ling Z, Pipeleers D, Gremlich S, Thorens B, Schuit F. Expression and functional activity of glucagon, glucagon-like peptide 1 and glucose-dependent insulinotropic peptide receptors in rat pancreatic islet cells. *Diabetes* 45: 257–261, 1996.

[25] Magnoni LJ, Míguez JM, Soengas JL. Glucagon effects on brain carbohydrate and ketone body metabolism of rainbow trout. *J. Exp. Zool.* 2001 Nov 1;290(6):662-71.

[26] Drucker DJ, Asa S. Glucagon gene expression in vertebrate brain. *J. Biol. Chem.* 1988 Sep 25;263(27):13475-8.

[27] Cockram CS, Kum W, Ho SK, Zhu SO, Young JD. Binding and action of glucagon in cultured mouse astrocytes. *Glia* 13: 141–146, 1995.

[28] Hoosein NM, Gurd RS. Identification of glucagon receptors in rat brain. *Proc. Natl. Acad. Sci. USA* 81: 4368–4372, 1984.

[29] Geary, N. 1990 Pancreatic glucagon signals postprandial satiety. *Neurosci. Biobehav. Rev.* 14, 323–338.

[30] Geary, N. 1998 Glucagon and the control of meal size. In Satiation. From gut to brain (ed. G. P. Smith), pp. 164–197. New York, NY: Oxford University press.

[31] Geary, N., Kissileff, H. R., Pi-Sunyer, F. X. and Hinton, V. 1992 Individual, but not simultaneous, glucagon and cholecystokinin infusions inhibit feeding in men. *Am. J. Physiol.* 262, R975–R980Intravenous infusion of glucagon also decreasesmeal size in humans.

[32] Geary, N. 1998 Glucagon and the control of meal size. In Satiation. From gut to brain (ed. G. P. Smith), pp. 164–197.

[33] Langhans,W., Geary,N. and Scharrer, E. 1982a Liver glycogen content decreases during meals in rats. *Am. J. Physiol.* 243, R450–R453.

[34] Geary, N. and Smith, G. P. 1983 Selective hepatic vagotomy blocks pancreatric glucagon's satiety effect. *Physiol. Behav.*31, 391–394].

[35] MacIsaac, L. and Geary, N. 1985 Partial liver denervations dissociate the inhibitory effects of pancreatic glucagon and epinephrine on feeding. *Physiol. Behav*. 35, 233–237.

[36] Arafat AM, Perschel FH, Otto B, Weickert MO, Rochlitz H, Schofl C, Spranger J, Mohlig M, Pfeiffer AF. Glucagon suppression of ghrelin secretion is exerted at hypothalamus-pituitary level. *J. Clin. Endocrinol. Metab.* 91: 3528–3533, 2006.

[37] Wynick, D., Hammond, P. J. and Bloom, S. R. 1993. The glucagonoma syndrome. *Clin. Dermatol.* 11, 93–97.

[38] Madsen, O. D., Karlsen, C., Blume, N., Jensen, H. I., Larsson, L. I. and Holst, J. J. 1995 Transplantable glucagonomas derived from pluripotent rat islet tumor tissue cause severe anorexia and adipsia. *Scand. J. Clin. Lab Invest. Suppl.* 220, 27–35 1993; Madsen et al. 1995.

[39] Wasserman DH, Spalding JA, Lacy DB et al. Glucagon is a primary controller of hepatic glycogenolysis and gluconeogenesis during muscular work. *Am. J. Physiol.* 1989; 257: E108–E117.

[40] Gerich JE, Langlois M, Noacco C, Karam J, and Forsham PH. Lack of a glucagon response to hypoglycemia in diabetes: evidence for an intrinsic pancreatic alpha-cell defect. *Science* 182: 171–173, 1973.

[41] Bessey PQ, Lowe KA. 1993. Early hormonal changes affect the catabolic response to trauma. *Ann. Surg.* 218:476–89.

[42] Cely CM, Arora P, Quartin AA, Kett DH, Schein RM. 2004. Relationship of baseline glucose homeostasis to hyperglycemia during medical critical illness. *Chest* 126:879–87.

[43] Fong YM, Marano MA, Moldawer LL,Wei H, Calvano SE, et al. 1990. The acute splanchnic and peripheral tissue metabolic response to endotoxin in humans. *J. Clin. Invest*. 85:1896–904.

[44] Stevenson RW, Steiner KE, Davis MA, Hendrick GK, Williams PE, et al. 1987. Similar dose responsiveness of hepatic glycogenolysis and gluconeogenesis to glucagon in vivo. *Diabetes* 36:382–89.

[45] Meinz HM, Lacy DB, Ejiofor J, McGuinness OP. 1998. Alterations in hepatic gluconeogenic amino acid uptake and gluconeogenesis in the endotoxin-treated conscious dog. *Shock* 9:296–303.

[46] Lang CH, Dobrescu C. 1989. In vivo insulin resistance during nonlethal hypermetabolic sepsis. *Circ. Shock* 28:165–78.

[47] McGuinness OP. 1994. The impact of infection on gluconeogenesis in the conscious dog. *Shock* 2:336–43.

[48] Holste LC, Connolly CC, Moore MC, Neal DW, Cherrington AD. 1997. Physiological changes in circulating glucagon alter hepatic glucose disposition during portal glucose delivery. *Am. J. Physiol*. 273:E488–96.

[49] Olsen, H.L. et al. (2005) Glucose stimulates glucagon release in single rat alpha-cells by mechanisms that mirror the stimulus-secretion coupling in beta-cells. *Endocrinology* 146, 4861–4870.

[50] Franklin, I. et al. (2005) Beta-cell secretory products activate alpha-cell ATP-dependent potassium channels to inhibit glucagon release. *Diabetes* 54, 1808–1815.

[51] Cabrera O, Berman DM, Kenyon NS, Ricordi C, Berggren PO, Caicedo A 2006 The unique cytoarchitecture of human pancreatic islets has implications for islet cell function. *Proc. Natl. Acad. Sci. USA* 103:2334–2339.

[52] Banarer S, McGregor VP, Cryer PE 2002 Intraislet hyperinsulinemia prevents the glucagon response to hypoglycemia despite an intact autonomic response. *Diabetes* 51:958–965.

[53] Ravier MA, Rutter GA 2005 Glucose or insulin, but not zinc ions, inhibit glucagon secretion from mouse pancreatic α-cells. *Diabetes*54:1789–1797.

[54] Samols E, Stagner JI, Ewart RB, Marks V: The order of islet microvascular cellular perfusion is β-α-δ in the perfused rat pancreas. *J. Clin. Invest.* 1988;82:350–353.

[55] Greenbaum, Havel P.J , Taborsky Jr. G.J. and Klaff L.J. Intra-islet insulin permits glucose to directly suppress pancreatic A cell function, *J. Clin. Invest.* 88 (1991), pp. 767–773.

[56] Franklin I, Gromada J, Gjinovci A, Theander S, Wollheim CB 2005. β-Cell secretory products activate α-cell ATP-dependent potassium channels to inhibit glucagon release. *Diabetes* 54:1808–1815.

[57] Xu E, Kumar M, Zhang Y, Ju W, Obata T, Zhang N, Liu S, Wendt A, Deng S, Ebina Y, Wheeler MB, Braun M, Wang Q 2006 Intraislet insulin suppresses glucagon release via GABA-GABAA receptor system. *Cell Metab*. 3:47–58.

[58] Gromada J, Ma X, Hoy M, Bokvist K, Salehi A, Berggren PO and Rorsman P 2004 ATP-sensitive KC channel-dependent regulation of glucagon release and electrical activity by glucose in wild-type and SUR1K/K mouse alpha-cells. *Diabetes* 53 (Suppl 3) S181–S189.

[59] Franklin I, Gromada J, Gjinovci A, Theander S and Wollheim CB 2005 b-cell secretory products activate a-cell ATP-dependent potassium channels to inhibit glucagon release. *Diabetes* 54 1808–1815.

[60] MacDonald PE, De Marinis YZ, Ramracheya R, Salehi A, Ma X, Johnson PR, Cox R, Eliasson L and Rorsman P. 2007 A K ATP channel-dependent pathway within alpha cells regulates glucagon release from both rodent and human islets of Langerhans. *PLoS Biology* 5 e143.

[61] Salehi A, Vieira E, Gylfe E 2006 Paradoxical stimulation of glucagon secretion by high glucose concentrations. *Diabetes* 55:2318–2323.

[62] Dunning, B.E. et al. (2005) Alpha cell function in health and disease: influence of glucagon-like peptide-1. *Diabetologia* 48,1700–1713.

[63] Schwanstecher, C. and Schwanstecher, M. (2002) Nucleotide sensitivity of pancreatic ATP-sensitive potassium channels and type 2 diabetes. *Diabetes* 51 (Suppl 3), S358–S362.

[64] Schwanstecher, C. et al. (2002) K(IR)6.2 polymorphism predisposes to type 2 diabetes by inducing overactivity of pancreatic beta-cell ATPsensitive K(+) channels. *Diabetes* 51, 875–879.

[65] Wiederkehr, A. and Wollheim, C.B. (2006) Minireview: implication of mitochondria in insulin secretion and action. *Endocrinology* 147, 2643–2649.

[66] Slucca M, Jamie S. Harmon J S, Oseid EA, Bryan J and Robertson RP. ATP-Sensitive K^+ Channel Mediates the Zinc Switch-Off Signal for Glucagon Response During Glucose Deprivation (2010). *DIABETES,* VOL. 59:128-134.

[67] Wendt A, Birnir B, Buschard K, Gromada J, Salehi A, Sewing S,Rorsman P, BraunM2004 Glucose inhibition of glucagon secretion from rat α-cells is mediated by GABA released from neighboring β-cells. *Diabetes* 53:1038–1045.

[68] Kieffer TJ, Habener JF (1999) The glucagon-like peptides. *Endocr. Rev.* 20:876–913.

[69] Eissele R, Göke R, Willemer S et al. (1992) Glucagon-like peptide-1 cells in the gastrointestinal tract and pancreas of rat, pig and man. *Eur. J. Clin. Invest.* 22:283–291.

[70] Kreymann B, Williams G, Ghatei MA, Bloom SR (1987) Glucagon-like peptide-1 [7-36]: a physiological incretin in man. *Lancet* 2:1300–1304.

[71] Ørskov C, Knuhtsen S, Baldissera FG, Poulsen SS, Nielsen OV, Holst JJ (1986) Glucagon-like peptides GLP-1 and GLP-2, predicted products of the glucagon gene, are secreted separately from pig small intestine but not pancreas. *Endocrinology* 119:1467–1475.

[72] Nauck M, Stöckmann F, Ebert R, Creutzfeldt W (1986) Reduced incretin effect in type 2 (non-insulin-dependent) diabetes. *Diabetologia* 29:46–54.

[73] Nauck MA, El-Ouaghlidi A, Gabrys B et al. (2004) Secretion of incretin hormones (GIP and GLP-1) and incretin effect after oral glucose in first-degree relatives of patients with type 2 diabetes. *Regul. Pept.* 122:209–217.

[74] Ørskov C, Jeppesen J, Madsbad S, Holst JJ (1991) Proglucagon products in plasma of noninsulin-dependent diabetics and nondiabetic controls in the fasting state and after oral glucose and intravenous arginine. *J. Clin. Invest.* 87:415–423.

[75] D'Alessio D, Thirlby R, Laschansky E, Zebroski H, Ensinck J (1993) Response of tGLP-1 to nutrients in humans. *Digestion* 54:377–379.

[76] Rocca AS, Brubaker PL. Role of the vagus nerve in mediating proximal nutrient-induced glucagon-like peptide-1 secretion. *Endocrinology* 1999; 140:1687-1694.

[77] Dunning BE, Foley JE, Ahren B 2005 α-Cell function in health and disease: influence of glucagon-like peptide-1. *Diabetologia* 48:1700–1713.

[78] Gromada J, Ma X, Hoy M, Bokvist K, Salehi A, Berggren PO, Rorsman P 2004 ATP-sensitive K_ channel-dependent regulation of glucagon release and electrical activity by glucose in wild-type and SUR1_/_ mouse _-cells. *Diabetes* 53(Suppl 3):S181–S189.

[79] Li Y, Cao X, Li LX, Brubaker PL, Edlund H, Drucker DJ 2005 β-Cell Pdx1 expression is essential for the glucoregulatory, proliferative,and cytoprotective actions of glucagon-like peptide-1. *Diabetes* 54:482–49.

[80] Guenifi A, Ahrén B, Abdel-Halim SM (2001) Differential effects of glucagon-like peptide-1 (7–36)amide versus cholecystokinin on

arginine-induced 40-Yamato E, Noma Y, Tahara Y et al. (1990) Suppression of synthesis and release of glucagon by glucagon-like peptide-1(7–36 amide) without affect on mRNA level in isolated rat islets. *Biochem. Biophys. Res. Commun.* 167:431–437.

[81] Ørskov C, Holst JJ, Nielsen OV (1988) Effect of truncated glucagon-like peptide-1 [proglucagon-(78–107) amide] on endocrine secretion from pig pancreas, antrum, and nonantral stomach.*Endocrinology* 123:2009–2013.

[82] Van Dijk G, Lindskog S, Holst JJ, Steffens AB, Ahrén B (1996) Effects of glucagon-like peptide-I on glucose turnover in rats. *Am. J. Physiol.* 270:E1015–E1021.

[83] Bishop AE, Polak JM: The anatomy, organization and ultrastructure of the islets of Langerhans. In Textbook of Diabetes. Pickup JC, Williams G, Eds. Oxford, U.K., *Blackwell Science,* 2002 , p.10.1–10.6.

[84] Reichlin S: Somatostatin. *N. Engl. J. Med.* 309:1495–1501, 1983.

[85] Reichlin S: Somatostatin (second of two parts). *N. Engl. J. Med.* 309:1556–1563, 1983.

[86] Gerich JE, Lorenzi M, Schneider V, Kwan CW, Karam JH, Guillemin R, Forsham PH: Inhibition of pancreatic glucagon responses to arginine by somatostatin in normal man and in insulin-dependent diabetics. *Diabetes* 23:876–880, 1974.

[87] Barden N, Lavoie M, Dupont A, Cote J, Cote JP: Stimulation of glucagon release by addition of anti-stomatostatin serum to islet of Langerhans in vitro. *Endocrinology* 101:635– 638, 1977.

[88] Ludvigsen E, Olsson R, Stridsberg M, Janson ET, Sandler S: Expression and distribution of somatostatin receptor subtypes in the pancreatic islets of mice and rats. *J. Histochem. Cytochem.* 52:391– 400, 2004.

[89] Strowski MZ, Parmar RM, Blake AD, Schaeffer JM: Somatostatin inhibits insulin and glucagon secretion via two receptors subtypes: an in vitro study of pancreatic islets from somatostatin receptor 2 knockout mice. *Endocrinology* 141:111–117, 2000.

[90] Hauge-Evans AC, King AJ, Carmignac D, Richardson CC, Robinson ICAF, Low MJ, Christie MR, Persaud SJ, Jones PM: Somatostatin secreted by islet glucose δ-cells fulfills multiple roles as a paracrine regulator of islet function. *Diabetes* 58:403– 411, 2009.

[91] Yoshimoto Y, Fukuyama Y, Horio Y, Inanobe A, Gotoh Mand Kurachi Y 1999. Somatostatin induces hyperpolarization in pancreatic islet [alpha] cells by activating a G protein-gated KC channel. *FEBS Letters* 444 265–269.

[92] Gromada J, Hoy M, Buschard K, Salehi A and Rorsman P 2001 Somatostatin inhibits exocytosis in rat pancreatic a-cells by Gi2-dependent activation of calcineurin and depriming of secretory granules. *Journal of Physiology* 535 519–532.

[93] Astrid C. Hauge-Evans, Aileen J. King, Danielle Carmignac, Carolyn C. Richardson, Iain C.A.F. Robinson, Malcolm J. Low, Michael R. Christie, Shanta J. Persaud,and Peter M. Jones: Somatostatin Secreted by Islet δ-Cells Fulfills Multiple Roles as a Paracrine Regulator of Islet Function. Diabetes, 58:403-411. February 2009.

[94] Cabrera O, Jacques-Silva MC, Speier S, Yang SN, Köhler M, Fachado A, Vieira E, Zierath JR, Kibbey R, Berman DM, Kenyon NS, Ricordi C, Caicedo A, Berggren PO.: Glutamate Is a Positive Autocrine Signal for Glucagon Release. *Cell Metab.* 2008 Jun;7(6):545-54.

[95] Bertrand, G., Gross, R., Puech, R., Loubatieres-Mariani, M.M., and Bockaert, J. (1993).Glutamate stimulates glucagon secretion via an excitatory amino acid receptor of the AMPA subtype in rat pancreas. *Eur. J. Pharmacol.* 237, 45–50.

[96] Hayashi, M., Otsuka, M., Morimoto, R., Hirota, S., Yatsushiro, S., Takeda, J., Yamamoto, A., and Moriyama, Y. (2001). Differentiation-associated Na+-dependent inorganic phosphate cotransporter (DNPI) is a vesicular glutamate transporter in endocrine glutamatergic systems. *J. Biol. Chem.* 276, 43400–43406.

[97] Uehara, S., Muroyama, A., Echigo, N., Morimoto, R., Otsuka, M., Yatsushiro, S., and Moriyama, Y. (2004). Metabotropic Glutamate Receptor Type 4 Is Involved in Autoinhibitory Cascade for Glucagon Secretion by α-Cells of Islet of Langerhans *Diabetes* 53, 998–1006.

[98] Date Y, Nakazato M, Hashiguchi S, et al. Ghrelin is present in pancreatic alpha-cells of humans and rats and stimulates insulin secretion. *Diabetes* 2002; 51:124–129.

[99] Volante M, Allia E, Gugliotta P, et al. Expression of ghrelin and of the GH secretagogue receptor by pancreatic islet cells and related endocrine tumors. *J. Clin. Endocrinol. Metab.* 2002; 87:1300–1308.

[100] Wierup N, Svensson H, Mulder H, Sundler F. The ghrelin cell: a novel developmentally regulated islet cell in the human pancreas. *Regul. Pept.* 2002; 107:63–69.

[101] Kojima M, Hosoda H, Date Y, et al. Ghrelin is a growth-hormone-releasing acylated peptide from stomach. *Nature* 1999; 402:656–660.

[102] Hosoda H, Kojima M, Mizushima T, et al. Structural divergence of human ghrelin. Identification of multiple ghrelin-derived molecules

produced by posttranslational processing. *J. Biol. Chem.* 2003; 278:64–70.

[103] Wren AM, Small CJ, Ward HL, et al. The novel hypothalamic peptide ghrelin stimulates food intake and growth hormone secretion. *Endocrinology* 2000; 141:4325–4328.

[104] Nakazato M, Murakami N, Date Y, et al. A role for ghrelin in the central regulation of feeding. *Nature* 2001; 409:194–198.

[105] Kojima M, Hosoda H, Date Y, et al. Ghrelin is a growth-hormone-releasing acylated peptide from stomach. *Nature* 1999; 402:656–660.

[106] Chuang JC, Sakata I, Kohno D, Perello M, Osborne-Lawrence S, Repa JJ, Zigman JM. Ghrelin Directly Stimulates Glucagon Secretion from Pancreatic {alpha}-*Cells. Mol. Endocrinol.* 2011 Jun 30. [Epub ahead of print].

[107] Dezaki K, Hosoda H, Kakei M, et al. Endogenous ghrelin in pancreatic islets restricts insulin release by attenuating Ca2þ signaling in beta-cells: implication in the glycemic control in rodents. *Diabetes* 2004; 53:3142–3151.

[108] Salehi A, Dornonville de la Cour C, Hakanson R, Lundquist I. Effects of ghrelin on insulin and glucagon secretion: a study of isolated pancreatic islets and intact mice. *Regul. Pept.* 2004; 118:143–150.

[109] Dezaki K, Hosoda H, Kakei M, et al. Endogenous ghrelin in pancreatic islets restricts insulin release by attenuating Ca2þ signaling in beta-cells: implication in the glycemic control in rodents. *Diabetes* 2004; 53:3142–3151.

[110] Ma X, Zhang Y, Gromada J, Sewing S, Berggren PO, Buschard K, Salehi A, Vikman J, Rorsman P, Eliasson L.: Glucagon stimulates exocytosis in mouse and rat pancreatic alpha-cells by binding to glucagon receptors. *Mol. Endocrinol.* 2005 Jan;19(1):198-212.

[111] Hartter E, Svoboda T, Ludvik B, Schuller M, Lell B, Kuenburg E et al. Basal and stimulated plasma levels of pancreatic amylin indicate its co-secretion with insulin in humans. *Diabetologia* 1991;34:52–4.

[112] Daniela Furrer, K. Kaufmann Claudia, E. Reusch and Thomas A. Lutz: Amylin reduces plasma glucagon concentration in cats. *The Veterinary Journal* Volume 184, Issue 2, May 2010, Pages 236-240.

[113] Nyholm B, Orskov L, Hove KY, Gravholt CH, Moller N, Alberti KG et al. The amylin analog pramlintide improves glycemic control and reduces postprandial glucagon concentrations in patients with type 1 diabetes mellitus. *Metabolism* 1999;48:935–41.

[114] Cryer PE. The prevention and correction of hypoglycemia. In Jefferson LS, Cherrington AD, eds. Handbook of Physiology. Section 7. The Endocrine System. Vol. II. The endocrine pancreas and regulation of metabolism. New York; Oxford University Press, 2001:1057.

[115] Esterhuizen AC, Spriggs TLB, Lever JD (1968) Nature of islet-cell innervation in the cat pancreas. *Diabetes* 17: 33±36.

[116] Orci L, Perrelet A, Ravazzola M, Malaisse-Lagae F, Renold AE (1973) A specialized membrane junction between nerve endings and B cells in islets of Langerhans. *Eur. J. Clin. Invest.* 3: 443±445.

[117] Bloom SR, Edwards AV (1981) Pancreatic endocrine responses to stimulation of the peripheral ends of the vagus nerve in conscious calves. *J. Physiol.* (Lond) 315: 31-41.

[118] Holst JJ, Grùnholt R, Schaffalitzky de Muckadell OB, Fahrenkrug J (1981) Nervous control of pancreatic endocrine secretion in pigs. 2. The effect of pharmacological blocking agents on the response to vagal stimulation. *Acta. Physiol. Scand.* 111: 9-14.

[119] Teff KL, Engelman K (1996) Oral sensory stimulation improves glucose tolerance in humans: effects on insulin, Cpeptide,and glucagon. *Am. J. Physiol.* 270:R1371-R1379.

[120] Ahren B. Autonomic regulation of islet hormone secretion–implications for health and disease. *Diabetologia* 2000; 43: 393–410.

[121] AhrØn B, Paquette TL, Taborsky GJ Jr (1986) Effect and mechanism of vagal nerve stimulation on somatostatin secretion in dogs. *Am. J. Physiol*. 250:E212-E217.

[122] Iversen J (1973) Effect of acetylcholine on the secretion of glucagon and insulin from the isolated, perfused canine pancreas. *Diabetes* 22: 381-387.

[123] Karlsson, AhrØn B (1993) Muscarinic receptor subtypes in carbachol-stimulated insulin and glucagon secretion in the mouse. *J. Auton. Pharmacol.* 13: 439-446.

[124] Havel PJ, Taborsky GJ Jr (1989) The contribution of the autonomic nervous system to changes of glucagon and insulin secretion during hypoglycemia stress. *Endocr. Rev*.10: 332-350.

[125] Karlsson S, AhrØn B (1991) Contribution of adrenergic nerves and the adrenals to 2-deoxy-D-glucose-induced insulin and glucagon secretion in the mouse. *Int. J. Pancreatol*. 10: 207-215.

[126] Evans ML, McCrimmon RJ, Flanagan DE, Keshavarz T, Fan X, McNay EC, Jacob RJ, Sherwin RS 2004 Hypothalamic ATP-sensitive K^+ channels play a key role in sensing hypoglycemia and triggering

counterregulatory epinephrine and glucagon responses. *Diabetes* 53:2542–2551.

[127] Borg MA, Sherwin RS, Borg WP, Tamborlane WV, Shulman GI 1997 Local ventromedial hypothalamus glucose perfusion blocks counterregulation during systemic hypoglycemia in awake rats. *J. Clin. Invest.* 99:361–365.

[128] Borg MA, Tamborlane WV, Shulman GI, Sherwin RS 2003 Local lactate perfusion of the ventromedial hypothalamus suppresses hypoglycemic counterregulation. *Diabetes* 52:663–666.

[129] McCrimmon RJ, Fan X, Ding Y, Zhu W, Jacob RJ, Sherwin RS 2004 Potential role for AMP-activated protein kinase in hypoglycemia sensing in the ventromedial hypothalamus. *Diabetes* 53:1953–1958.

[130] Miki T, Liss B, Minami K, Shiuchi T, Saraya A, Kashima Y, Horiuchi M, Ashcroft F, Minokoshi Y, Roeper J, Seino S 2001. ATP-sensitive K^+ channels in the hypothalamus are essential for the maintenance of glucose homeostasis. *Nat. Neurosci.* 4:507–512.

[131] de Jong, A., Strubbe, J. H. and Steffens, A. B. 1977 Hypothalamic influence on insulin and glucagon release in the rat. *Am. J. Physiol.* 233, E380–E388.

[132] Secchi, A., Caldara, R., Caumo, A., Monti, L. D., Bonfatti, D., Di Carlo, V. and Pozza, G. 1995 Cephalic-phase insulin and glucagon release in normal subjects and in patients receiving pancreas transplantation. *Metabolism* 44, 1153–1158.

[133] Langhans,W., Pantel, K., Muller-Schell,W., Eggenberger, E. and Scharrer, E. 1984 Hepatic handling of pancreatic glucagon and glucose during meals in rats. *Am. J. Physiol.* 247, R827–R832.

[134] Woods SC, Lutz TA, Geary N, Langhans W. Pancreatic signals controlling food intake; insulin, glucagon and amylin. *Phil. Trans. R. Soc. B* (2006) 361, 1219–1235.

[135] Cryer PE. Glucagon and hyperglycaemia in diabetes. *Clin. Sci.* (Lond) 2008;114: 589–590.

[136] Unger RH, Orci L. Paracrinology of islets and the paracrinopathy of diabetes. *Proc. Natl. Acad. Sci. USA* 2010; 107: 16009–16012.

[137] Cryer PE, Davis SN, Shamoon H. Hypoglycemia in diabetes. *Diabetes Care* 2003; 26: 1902–1912.

[138] Ohneda A, Watanabe K, Horigome K et al. (1978). Abnormal response of pancreatic glucagon to glycemic changes in diabetes mellitus. *J. Clin. Endocrinol. Metab.* 46:504–510.

[139] Dunning BE, Gerich JE. The role of alpha-cell dysregulation in fasting and postprandial hyperglycemia in type 2 diabetes and therapeutic implications. *Endocr. Rev.* 2007; 28: 253–283.

[140] Unger RH, Aguilar-Parada E, Muller WA, Eisentraut AM. Studies of pancreatic alpha cell function in normal and diabetic subjects. *J. Clin. Invest.* 1970; 49: 837–848.

[141] Unger RH. Glucagon physiology and pathophysiology in the light of new advances. *Diabetologia* 1985; 28: 574–578.

[142] Dinneen S, Alzaid A, Turk D, Rizza R 1995 Failure of glucagon suppression contributes to postprandial hyperglycemia in IDDM. *Diabetologia* 38:337–343.

[143] Shah P, Basu A, Basu R, Rizza RA 1999 Impact of lack of suppression of glucagon on glucose tolerance in humans.*Am. J. Physiol.* 277:E283–E290.

[144] Shah P, Vella A, Basu A, Basu R, Schwenk WF, Rizza RA 2000 Lack of suppression of glucacon contributes to postprandial hyperglycemia in subjects with type 2 diabetes mellitus. *J. Clin. Endocrinol. Metab.* 85:4053–4059.

[145] Hamaguchi T, Fukushima H, Uehara M, Wada S, Shirotani T, Kishikawa H, Ichinose K, Yamaguchi K, Shichiri M 1991 Abnormal glucagon response to arginine and its normalization in obese hyperinsulinaemic patients with glucose intolerance: importance of insulin action on pancreatic α-cells. *Diabetologia* 34:801–806.

[146] J. C. Henquin and J. Rahier 2011. Pancreatic alpha cell mass in European subjects. *Diabetologia* 54:1720–1725.

[147] Meier JJ, Kjems LL, Veldhuis JD, Lefebvre P, Butler PC 2006 Postprandial suppression of glucagon secretion depends on intact pulsatile insulin secretion: further evidence for the intraislet insulin hypothesis. *Diabetes* 55:1051–1056.

[148] Parker JC, Andrews KM, Allen MR, Stock JL, McNeish JD 2002 Glycemic control in mice with targeted disruption of the glucagon receptor gene. *Biochem. Biophys. Res. Commun.* 290:839–843.

[149] Gerich JE, Langlois M, Noacco C, Karam JH, Forsham PH 1973. Lack of glucagon response to hypoglycemia in diabetes: evidence for an intrinsic pancreatic α-cell defect. *Science* 182:171–173.

[150] Bolli GB, Tsalikian E, Haymond MW, Cryer PE, Gerich JE 1984 Defective glucose counterregulation after subcutaneous insulin in noninsulin-dependent diabetes mellitus: paradoxical suppression of glucose utilization and lack of compensatory increase in glucose

production, roles of insulin resistance, abnormal neuroendocrine responses, and islet paracrine interactions. *J. Clin. Invest.* 73:1532–1541

[151] Campbell LV, Kraegen EW, Meler H, Lazarus L 1979 Hormonal responses to insulin infusion in diabetes mellitus. *Diabetologia* 16:359–364.

[152] Meneilly GS, Cheung E, Tuokko H 1994 Counterregulatory hormone responses to hypoglycemia in the elderly patient with diabetes. *Diabetes* 43:403–410].

[153] Shamoon H, Friedman S, Canton C, Zacharowicz L, Hu M, Rossetti L 1994 Increased epinephrine and skeletal muscle responses to hypoglycemia in non-insulin-dependent diabetes mellitus. *J. Clin. Invest.* 93:2562–2571].

[154] Larsson H, Berglund G, Ahre´n B 1995 Glucose modulation of insulin and glucagon secretion is altered in impaired glucose tolerance. *J. Clin. Endocrinol. Metab.* 80:1778–1782.

In: Glucagon
Editor: Ahmed Al-Jebawi
ISBN: 978-1-62618-147-2

Chapter 3

Physiological Effects of Glucagon on Cardiovascular System

M. A. Rosic, V. Zivkovic, S. Pantovic, M. Colic and V. Lj. Jakovljevic*

Department of Physiology, Faculty of Medical Sciences,
University of Kragujevac, Kragujevac, Serbia

Abstract

Glucagon is an endogenous polypeptide hormone that exerts both positive inotropic and the chronotropic effects on the myocardial tissue. Cardiac effects of glucagon are considered to be resultant from stimulation of glucagon receptors associated with Gs protein, which causes adenylyl cyclase (AC) activation and the consequent increase of 3',5',-cyclic adenosine monophosphate (cAMP) production. It is used extensively to treat beta-blockers or calcium channel blockers overdose. Interestingly, phosphodiesterase 3 inhibitors can enhance this well known positive inotropic effect of glucagon on ventricular myocardium. It is also noticed that glucagon induces increase in contractility only in ventricle, but not in atrium, which seems to be a consequence of a lower glucagon receptor density in this tissue. In order to elevate sinus rate above that of the dominant ventricular focus and thus allowing a return to dominance of the sinus node, the antiarrhytmic effect of glucagon in ventricular

* Email address: drvladakgbg@yahoo.com.

arrhythmias is mediated through a supraventricular action. Other findings showed that the cardiac anaphylactic crisis was markedly reduced in the presence of glucagon. Furthermore, these data suggested that the antiarrhythmic action of glucagon during the cardiac anaphylaxis involves inhibition of histamine release, vasodilation of the coronary vessels, stimulation of the automaticity of the sinoatrial node and enhancement of the atrioventricular conduction velocity. Results of our research group showed that the positive effect of glucagon on ischemic vasodilation is due to a faster restoration of the coronary perfusion pressure during the reperfusion period. In addition to its effects on the heart, there are data to indicate that glucagon has protective influence on impaired cerebrovasodilation by upregulating cAMP, suggesting that it may provide neuroprotection when given after traumatic brain injury, or prior to certain neurosurgical or cardiac interventions in which the incidence of perioperative ischemia is high.

The Effects of Glucagon on Cardiac Function

Data about the effects of glucagon on cardiac muscle are limited. This is corroborated by just a few experimental studies which investigated the role of glucagon in myocardial function. Cardiac effects of glucagon are considered to be resultant from stimulation of glucagon receptors associated with Gs protein, which causes adenylyl cyclase (AC) activation and the consequent increase of 3',5',-cyclic adenosine monophosphate (cAMP) production (White 1999). By activating this pathway, glucagon increases heart rate and contractility (White 1999). Consequently, it is used for the management of poisoning caused by cardiodepressant drugs such as β-adrenoceptors (β-AR) blockers or calcium channel blockers (DeWitt and Waksmann 2004). Recent findings have emerged to enable a better understanding of the impact of glucagon on the contraction of heart muscle. Namely, it has been demonstrated that glucagon (0.01-1 μM) produces a concentration-dependent positive inotropic effect in ventricular but not in atrial myocardium (Gonzalez-Muñoz 2008). The mechanism of this effect is still uncertain, but it seems to be a consequence of a lower glucagon receptor density in atrium tissue (Gonzalez-Muñoz 2008).

On the other hand, it has been recognized that glucagon possesses potent chronotropic and inotropic potential (Whitehouse 1966, Glick 1968, Farah 1960) similar to the well-known cardiac actions of the catecholamines. It has been also suggested that both action of glucagon and catecholamines are similar via stimulating adenylyl cyclase (Entman 1969, Epstein 1970). The actions of these hormones must, however, be mediated through different receptors, since beta-receptor blockade with propranolol has no effect on glucagon action (Lucchessi 1968). Several authors have reported that glucagon has no effect on ventricular automaticity (Steiner 1969, Lucchesi 1969, Cohn 1970), whereas it has long been accepted that catecholamines are potent stimulants of ventricular automaticity. Disagreement does exist concerning the effect of glucagon on this parameter. Lipski et al. (Lipski 1969), in a preliminary report, noted an increase in ventricular automaticity after glucagon administration in dogs with intact conduction systems. Daniell and coauthors (Daniell 1970) reported similar results in dogs with chronic and acute complete heart block. An increase in ventricular automaticity in dogs with acute complete heart block was noted for only 1 minute after glucagon administration. In the next report from this laboratory, glucagon was shown to enhance the automaticity of isolated canine Purkinje fibers, a strong indication that it increases ventricular automaticity of the heart *in situ* (Pruett JK 1971). In agreement with previous studies, Wilkerson and coworkers (Wilkerson 1971) also showed that glucagon increases ventricular automaticity, but this increase may be masked by a rate-dependent depression of ventricular automaticity effected by the potent positive chronotropic action of the hormone.

Positive effects of glucagon on the heart muscle still exist even in a presence of impaired cardiac function. High-bolus dose of glucagon (2 mg/kg) followed by glucagon infusion significantly increases heart rate, mean arterial pressure and reduces prolonged QRS interval in rat model of amitriptyline-induced cardiovascular toxicity (Kaplan 2008). This may be a very interesting field of investigation and a potential therapeutic use of glucagon in patients with heart failure caused by cardiotoxic drugs.

Glucagon could also affect the myocardial transport and utilization of energy-metabolites coming from the coronary microcirculation. Rosic and coworkers (Rosic et al, 1988) found that myocardial utilization of [3H]-or [14C]-labelled short-chain monocarboxylic acids (L-lactate, pyruvate and acetate) was significantly reduced by glucagon, to the same degree as their unidirectional sarcolemmal transport. Inhibition of L-[14C]lactate uptake was dose-dependent and in positive correlation with myocardial lactate production. It is concluded that glucagon may regulate sarcolemmal permeability and myocardial utilization for energy-metabolites from the coronary circulation (Rosic et al. 1988), suggesting a very important role of this hormone in supplying energy to the cardiomyocytes.

There are only few studies on the effects of glucagon on human heart. Most of these studies investigated the potential role of glucagon in the treatment of symptomatic bradycardia (Love 1997, Love 1998). It has been proven that intravenous administration of glucagon can significantly improve the clinical condition of patients with symptomatic bradycardia, when atropine failed to show an adequate effect (Love 1998). In addition, the same group of authors (Love 1997) showed that, when symptomatic bradycardia is not related to acute beta-blocker ingestion, glucagon may also have a positive effect and obviate the need for further therapy. This effect of glucagon certainly deserves future controlled studies in order to estimate the justification for its use in some emergent cardiovascular events (Love 1993).

The most recent data show that the combined use of phosphodiesterases' (PDEs) inhibitiors and glucagon may enhance cardiac automaticity (Gonzalez-Muñoz 2011). Taken into consideration the mechanism by which glucagon achieves its cardiac effects through stimulation of cAMP production and the fact that PDE inhibitors prevent degradation of cAMP, it is reasonable to assume that PDE inhibitors and glucagon produce a synergistic inotropic effect. This might have some practical interest since both glucagon and PDE inhibitors are considered to be useful agents for treating toxicity induced by cardiodepressant agents such as β-adrenoceptor blockers or calcium channels blockers (DeWitt and Waksmann 2004). Nevertheless, this mutual effect, by enhancing cardiac automaticity, increases the risk of cardiac arrhythmias (Gonzalez-Muñoz 2011). However, the clinical relevance of these findings remains to be determined.

The Effects of Glucagon on Ischemia/ Reperfusion Injury in the Heart

In order to understand the potential effects of glucagon on myocardial ischemia and the reperfusion, it is necessary to elucidate the molecular events during this period. The myocardial reperfusion following ischemia leads to the ischemic vasodilation through affecting the release of various vasoactive substances, such as free radicals (Bertuglia 2003), nitric oxide (NO) (Tsao 1990) and histamine (Asanuma 2006). In addition, while low doses of NO appear to play a key protective role from the myocardial ischemia-reperfusion injury (Masini 1991), high concentrations of NO exert a highly toxic and harmful effect during the ischemia-reperfusion injury (Jugdutt 2002). The stimulation of the H_2 receptors by the histamine induces both positive inotropic and positive chronotropic effects (Hough 2001). It has been shown that the blockage of the histamine H_2 receptors improved the anaerobic metabolism of the myocardium in the ischemic heart disease that has been associated with the reduced myocardial oxygen consumption demonstrating the highly beneficial effect against the ischemia-reperfusion injury. On the basis of this data, our research group (Rosic et al. 1987, 1988) demonstrated that glucagon may have a positive effect on ischemia/reperfusion injury by elevating the histamine transport in the isolated guinea pig heart.

On the other hand, glucagon can also have a beneficial role during cardiac anaphylaxis. Obtained results showed that the cardiac anaphylactic crisis was markedly reduced in the presence of glucagon (Andjelkovic and Zlokovic 1982). The antiarrhythmic action of glucagon during the cardiac anaphylaxis involved the inhibition of histamine release, the vasodilation of the coronary vessels, the increase in the automaticity of the sinoatrial node and the enhancement of the atrioventricular conduction velocity (Andjelkovic and Zlokovic 1982). Based on the observations from this experimental work, glucagon might have useful clinical applications in cardiac patients with failure and arrhythmias due to allergic immediate reactions.

It seems that glucagon can affect the changes of the coronary perfusion pressure (CPP) during the ischemia/reperfusion, by altering the release of NO and/or histamine in the heart. Therefore, in our recent study, we investigated the changes in the coronary perfusion pressure during the ischemia-reperfusion period, as well as the concentrations of nitric oxide, thiobarbituric acid reactive substances (TBARS), and histamine either in the presence or in the absence of glucagon in the coronary venous effluent of the isolated rat heart (Rosic et al, 2010). The summary of our results showed that in the presence of glucagon, there was a faster restoration of coronary perfusion pressure during ischemic vasodilation compared to the absence of glucagon with no apparent changes in TBARS concentration. The glucagon's administration leads to the decreased release of histamine by approximately 35%. The biphasic release of NO in the presence of glucagon initially showed an augmentation by 60% but later it was followed by the significant attenuation of 45%. In this study, glucagon increased NO release both during the preischemic period and within the first 30 seconds after the ischemic vasodilation, which resulted in stabilization of mast cells and prevented the histamine release during the ischemic vasodilation. On the other hand, the decreased histamine release attenuated stimulation of the endothelial histamine receptor (H_1) and as a consequence decreased release of NO, thus switched off this vasodilatory mechanism. Simultaneously, the decreased histamine release may lead to the attenuation of the H_2/H_3 mediated coronary vasodilation in the presence of glucagon. As a result, the duration of ischemic vasodilation is significantly shorter in the presence of glucagon. Furthermore, the increased NO release within the first 30 seconds of the ischemic vasodilation leads to glucagon associated vasodilation that in turn may also increase the rigidity of the coronary blood vessels, generating the faster development of CPP. It seems that even though the mechanism(s) of the glucagons influences on the coronary vessels during the ischemic vasodilation are still unclear, our results clearly indicate the key involvement of both histamine and NO during this process.

The Effects of Glucagon-Like Peptide-1 (GLP-1) on Myocardial Function

Glucagon-like peptide-1 (GLP-1) is a member of the proglucagon incretin family secreted by intestinal L-cells in response to nutrient ingestion (Baggio 2007, Persson et al. 2000, Lund 2005, Meier et al. 2004). It is produced in significantly smaller amounts by the pancreas and hypothalamus (Suda et al. 1989). GLP-1 is a product of the same gene which encodes glucagon, a fact that was proven after cloning it (Bell et al. 1983). GLP-1 has two major isoforms (7-36 amide and 7-37) which are among the most potent stimulators of glucose-dependent insulin secretion (Fehman et al. 1995). Enzymatic processing of preproglucagon into proglucagon results in active forms from which a major proglucagon fragment is then detached (Rouillé et al. 1995, Damholt et al. 1999). GLP-1 modulates multiple cellular functions believably via acting on GLP-1 receptors (GLP-1Rs) (Runge 2008, Campos et al. 1993, Drucker et al. 1987, 2006), which show a structural similarity to the glucagon receptor (Wilmen et al. 1998) and are expressed in the human pancreas, heart, lung, kidney, stomach and brain (Kavianipour 2003). It stimulates glucose-dependent insulin secretion and insulin biosynthesis and inhibits glucagon secretion, gastric emptying, and food intake (Barnett 2009, Ørskov et al. 1993).

Although the major physiological function of GLP-1 is associated with glycemic control, increasing evidence indicates that GLP-1 may also play an important role in cardiovascular physiology (Grieve 2009). GLP-1R knockout mice (GLP-1R-/-) have lower heart rate and elevated left ventricular (LV) end-diastolic pressure, and increased LV thickness (Ban 2008). In clinical studies, using liraglutide (Victoza™), a GLP-1R agonist, a consistent decrease in blood pressure and a reduction of cardiovascular risk markers were observed in a cohort of over 5000 patients (Madsbad 2009).

The effects of GLP-1 on myocardium may be potentially mediated via its G protein-coupled receptors (Holz 2008, Zhao 2006). Recently, a new glucagon-like peptide isolated from the intestine of the eel, *Anguilla japonica* (Uesaka 2001) with a structure similar to that of oxynto-modulins has shown an inotropic effect via stimulation of Ca^{2+} influx and a chronotropic effect independent of extracellular Ca^{2+}. However, the effects of GLP-1 on myocyte electrophysiology are still uncertain. The group of Xiao et al. (Xiao 2011) tried to elucidate this topic by investigating the effects of GLP-1 on the voltage-gated L-type Ca^{2+} channel in isolated canine left ventricular myocytes. They demonstrated for the first time that GLP-1 enhances voltage-gated L-type Ca^{2+} currents in canine cardiomyocytes via the cAMP-dependent protein kinase A mechanism and may contribute, at least partially, to the prolongation of the action potential duration. These effects are most likely mediated via the activation of the cardiac GLP-1R and then the cAMP-dependent PKA pathway. These findings suggest a novel mechanism of the potential beneficial cardiac effects of GLP-1R agonists on the improvement of myocardial contractility.

Although receptors for GLP-1 have been found in a variety of extrapancreatic tissues including the heart (Bullock 1996; Wei and Mojsov, 1996), direct cardiovascular actions of GLP-1 remain controversial. For example, GLP-1 increased heart rate and blood pressure in intact rodents through sympathostimulatory effects (Barragan 1994, 1996; Yamamoto 2002) but depressed myocardial contractility in isolated rat ventricular myocytes (Vila Petroff 2001). Recently, infusion of GLP-1 was associated with improvements in myocardial glucose uptake and left ventricular (LV) systolic function in conscious dogs with dilated cardiomyopathy but not in normal dogs (Nikolaidis 2004a).

As such, whether the actions of GLP-1 on myocardial function and glucose uptake are mediated by direct effects and whether they are evident in both normal and ischemic myocardium remains unresolved. The potential answer for this question can be found in research of Zhao and coauthors (Zhao 2006). They demonstrated that GLP-1 had direct negative effects on myocardial contractility but stimulated myocardial glucose uptake and lactate production in normally perfused, isolated rat hearts. Although the effects of GLP-1 on myocardial contractility were distinct from those of insulin, they were quite comparable to the effects of insulin on coronary flow and myocardial glucose uptake. Furthermore, the myocardial glucose uptake stimulatory effects of GLP-1 were not mediated through the classic insulin-signaling cascade involving Akt-1 activation and GLUT-4 translocation but rather through increased myocardial nitric oxide production, p38 MAP kinase activity, and GLUT-1 translocation (Zhao 2006). These findings provide new insights into the direct myocardial effects of GLP-1 and novel signaling mechanisms for glucose uptake, independent of the Akt-1 activation.

Nevertheless, more studies are needed to fully understand the role of GLP-1 on cardiac function and it`s potential use as adjuvant therapy in different cardiovascular events.

The Effects of GLP-1 on Ischemia/ Reperfusion Injury in the Heart

Potential beneficial effects of GLP-1 on cardiac ischemia/reperfusion injury can be based on the pharmacological properties of GLP-1 (insulinomimetic (D'Alessio 1994) and insulinotropic (Ritzel 1995) actions) as a means to stimulate myocardial glucose uptake during postischemic contractile dysfunction. However, there have been a limited number of studies examining the effects of GLP-1 on infarct size or recovery from a post-ischemic injury with conflicting results. Zhao et al. (Zhao 2006) have reported that GLP-1 increased myocardial glucose uptake and enhanced recovery of cardiac function after low-flow ischemia in postischemic isolated rat hearts. On the other hand, Kavianipour et al. (Kavianipour 2003) reported that GLP-1-(7–36) did not alter myocardial glucose utilization, hemodynamics, or infarct size in an acute porcine model of infarction, despite the fact that GLP-1 increased insulin secretion and decreased lactate uptake. In contrast, studies on canines has demonstrated that a 24-h infusion of GLP-1 (2.5 pmol/kg/min) mitigated postischemic contractile dysfunction and improved ventricular relaxation in conscious dogs after brief periods (10 min) of complete coronary artery occlusion.

Nikolaidis et al. demonstrated for the first time that GLP-1 infusion improved regional and global LV function in patients with acute myocardial infarction (AMI) and severe systolic dysfunction after successful primary angioplasty. Because the design of this study was limited to the first 72 hours after MI, we cannot speculate whether continued GLP-1 infusion could have conferred greater cardiovascular benefits. Even if the benefits were limited to accelerated recovery from early myocardial stunning, this would be clinically important in post-MI patients with severe LV dysfunction.

However, of specific interest is the fact that GLP-1 has also been shown to promote the activity of phosphoinositide 3-kinase (PI3K) in β-cells (Buteau 1999). This kinase is clearly associated with myocardial protection in the setting of ischemic/reperfusion injury (Hausenloy 2004) as well as myocardial preconditioning (Mocanu 2002). Recent data have suggested that GLP-1 can exert a direct cytoprotective effect via inhibition of apoptosis either directly in target cells expressing the GLP-1 receptor or possibly via the activation of survival factors (Redondo 2003). GLP-1 has also been shown to protect against apoptosis in insulinoma cell lines through both cAMP and PI3K (Hui 2003). In this context, it is important to bear in mind that insulin can activate prosurvival kinases such as PI3K (Jonassen 2000), which have been proposed as integral components of antiapoptotic cascades involved in myocardial protection. The group of Bose and coworkers (Bose 2005) has shown for the first time that GLP-1 protects against myocardial infarction in the isolated and intact rat heart. This protection appears to be mediated by PI3K, pro-survival kinase which leads to the phosphorylation and inactivation of the pro-apoptotic peptide BAD.

Investigations on mice model confirmed the importance of PI3K signal pathway in the mechanisms of GLP-1 cardioprotection. In this model, GLP-1 showed a direct protective action on cardiac myocytes from ischemic/reperfusion (I/R) induced apoptosis through PI3K/Akt and MAPK/ERK signaling pathways. Knowing potentially ideal antidiabetic properties of GLP-1 (Meneilly 2001), this finding may represent a new therapeutic potential for this class of drug currently undergoing clinical trials in the treatment of type 2 diabetes.

Thus, it could be clinically important to validate the cardioprotective effects of GLP-1 agonists and explore their novel protective mechanism in diabetic patients who are at a high risk of developing cardiovascular diseases. Albiglutide, a long lasting GLP-1 analog, is currently being investigated in a phase III clinical trial for its potential to treat type2 diabetes (St Onge 2010). Experiments on rats have shown that Albiglutide reduced myocardial infarct size and improved post-ischemic cardiac function energetics following myocardial I/R injury. The observed benefits were associated with enhanced myocardial glucose uptake and shift toward a more energetically favorable substrate metabolism profile by increasing both glucose and lactate oxidation (Bao 2011). These findings suggest that in addition to providing peripheral glycemic control, albiglutide may have a direct therapeutic potential for improving cardiac energetics and function in the setting of myocardial ischemic injury.

The Effects of Glucagon on Blood Vessels

Although the circulatory effects of glucagon have been investigated for over 50 years, precise mechanisms of action and physiological effects of glucagon on blood vessels have not been completely explained yet. It has been shown that glucagon is a potent vasoactive substance with a considerable variation in the response of the various regional vascular beds (Ross 1970). This polypeptide hormone mainly produces a reduction in mean arterial pressure and in total peripheral resistance, increases the blood flow and therefore works as a vasodilator (Ross 1970; Glick et al. 1968; Williams et al. 1969; Stuhlinger et al. 1974). Glucagon causes a fast-acting direct renal vasodilatation in rat in vivo and vasorelaxation in dog blood vessels in vitro (Okamura et al. 1986; Tolins 1992.). Removal of the endothelium from the artery did not alter the magnitude of relaxation induced by glucagon (Okamura et al. 1986, Ding et al. 2000). Moreover, dilatation has been demonstrated in denervated isolated canine intestinal loops (Varro and Czernay, 1966). Glucagon-induced vasodilatation is considered to be resultant from stimulation of glucagon receptors associated with G protein, which causes adenylyl cyclase activation and the consequent increase of 3`, 5`,-cyclic adenosine monophosphate (cAMP) content in vascular smooth muscle (Farah AE, 1983; Lok et al. 1994; Brubaker and Drucker 2002; Authier and Desbuquois 2008). Subsequently, cAMP-dependent phosphorylation of the myosin light chain kinase may occur (Farah AE, 1983). The myosin light chain kinase shows reduced sensitivity to the Ca^{2+}calmodulin complex when it is phosphorylated and thus may produce relaxation of smooth muscle (Farah AE, 1983). Variation in the response of blood vessels to glucagon may be due to different extents of production of cyclic AMP in smooth muscle cells, which may be the consequence of differences in the quantity and sensitivity of glucagon receptors in some blood vessels (Okamura et al. 1986; Komatsu et al. 1997). On the other hand, there is data that showed vasoconstrictive effects of glucagon on the hepatic arterial bed (Ross 1970).

In liver cells, glucagon activates multiple G protein-mediated signal transduction pathways not only via stimulation of adenylyl cyclase (Pohl et al. 1969) but also via stimulation of glucagon receptor signaling coupled to inositol phosphate production (Wakelam et al. 1986). Unfortunately, there is no information about this glucagon induced signal transduction pathways in the blood vessels.

Armstead et al. confirmed the vasodilator effects of glucagon in the brain, by glucagon-induced upregulation of cAMP, suggesting that it may provide neuroprotection when given after traumatic brain injury, or prior to certain neurosurgical or cardiac interventions in which the incidence of perioperative ischemia is high (Armstead et al. 2011). It has been noted that glucagon infused at a physiologic dose does not increase splanchnic blood flow, although it increases portal tributary blood flow at supraphysiological doses (Cerini et al. 1989).

Research on isolated human coronary endothelial cells (Ding et al. 2000) revealed another anti-insulin action of glucagon beyond its known effects on glucose metabolism. Long-term glucagon's influence on eNOS and NO production in isolated human coronary endothelial cells showed that glucagon, per se, has no discernible effect on either eNOS expression or NO production. However, glucagon abolished the stimulatory effect of insulin on eNOS expression and NO production in these cells. The mechanism of this inhibition is still unknown (Ding et al. 2000).

On the other hand, the above mentioned results of our study (Rosic et al, 2010) in which we investigated the influence of glucagon on ischemic vasodilatation of the isolated rat heart showed that NO level during preischemic period was significantly elevated in the presence of glucagon. Even though the NO level was significantly elevated, the amount of glucagon-induced NO release seemed to be insufficient to dilate coronary blood vessels, but sufficient to stabilize mast cells and consequently reduce the histamine release during ischemic vasodilatation. As a result, duration of the ischemic vasodilatation is significantly shorter in the presence of glucagon, i.e. restoration of the coronary perfusion pressure during the reperfusion period is faster (Rosic et al, 2010).

The Effects of GLP-1 on Blood Vessels

Apart from the above mentioned influence of glucagon on blood vessels, it has been recently shown that GLP-1 has a great impact on properties of blood vessels.

The first observations concerning GLP -1's vasodilatatory effect via an endothelium - dependent mechanism were made in the pulmonary artery's vascular bed (Richter et al. 1993). Recently, it has been observed that a product of GLP -1(7-36) degradation by depeptidyl peptidase-4 (DPP -4), namely GLP -1(9-36), also has a vasodilatatory effect, and that this activity does not require the GLP -1R (Ban et al. 2008).

Recently, many researchers have turned their attention to the cardiovascular effects of GLP-1 and have found that GLP-1 can induce endothelium-dependent vasorelaxation (Golpon et al. 2001, Nathanson et al. 2009, Ban et al. 2008) and improve the response of vessels to other vasodilators (Yu et al. 2003, Nystrőm et al. 2004, Basu et al. 2007). Furthermore, GLP-1 can elevate the NO levels in coronary effluent from mouse hearts, and its vasorelaxing effect can be blocked by an eNOS inhibitor (Ban et al. 2008), suggesting that GLP-1 upregulates eNOS.

GLP-1 exerts its actions on the blood vessels through GLP-1 receptor (GLP-1R), expressed in the endothelium (Ban et al. 2008, Nystrom et al. 2004, Ishibashi et al. 2010). Native GLP-1, existing mainly as an intact, biologically active form (GLP-1 [7–36 amide]), is rapidly degraded after secretion by DPP-4 to its inactive form, N-terminally truncated GLP-1(9–36). GLP-1(9–36) has a weak affinity to GLP-1R and was previously considered to have no biological function. However, some research has suggested that GLP-1(9–36), which offers significant cardioprotection against ischemia-reperfusion injury and induces vasodilation (Nathanson et al. 2009, Ban et al. 2008, Ban et al. 2010), may affect GLP-1 function, at least in the cardiovascular system.

Ding et al. showed that the incubation of human umbilical vein endothelial cells (HUVECs) with GLP-1 increased eNOS activity which is dose dependent (Ding et al. 2012). They also showed that GLP-1-induced NO production was blocked by N G-nitro-L-arginine methyl ester - L-NAME (an eNOS inhibitor), suggesting that eNOS activation mediates the effect of GLP-1 on NO release in HUVECs. Previous research has suggested that GLP-1 can also induce pulmonary vasodilation in isolated rat lungs that can be blocked by L-NAME (Golpon et al. 2001).

Conclusion

Although the basic physiological effects of glucagon and its derivatives are well known for many years, more research is needed to explore the effects of this hormone on the human body in general and the cardiovascular system in particular. At the present time, it looks like glucagon and its derivatives are potentially useful drugs for the treatment various cardiovascular diseases.

Fortunately, the research in this field is rapidly progressing and may one day revolutionize the treatment of cardiovascular diseases.

References

Andjelkovic I and Zlokovic B. "Protective effects of glucagons during the anaphylactic response in guinea-pig isolated heart,"*British Journal of Pharmacology*, vol. 76, no. 3, pp. 483–489,1982.

Armstead WM, Kiessling JW, Cines DB, Higazi AA. Glucagon protects against impaired NMDA-mediated cerebrovasodilation and cerebral autoregulation during hypotension after brain injury by activating cAMP protein kinase A and inhibiting upregulation of tPA. *J. Neurotrauma* 28(3): 451-457; 2011.

Asanuma H, Minamino T, Ogai A, Kim J, Asakura M, Komamura K, Sanada S, Fujita M, Hirata A, Wakeno M, Tsukamoto O, Shinozaki Y, Myoishi M, Takashima S, Tomoike H, Kitakaze M. "Blokade of histamine H2 receptors protects the heart against ischemia and reperfusion injury in dogs," *Journal of Molecular and Cellular Cardiology,* vol. 40, no. 5, pp. 666–674, 2006.

Authier F and Desbuquois B. Glucagon receptors. *Cell Mol. Life Sci.* 65: 1880-1899, 2008.

Baggio LL, Drucker DJ (2007) Biology of incretins GLP-1 and GIP. Gastroenterology 132: 2131–2157.

Ban K, Kim KH, Cho CK, Sauvé M, Diamandis EP, Backx PH, Drucker DJ and Husain M. Glucagon-like peptide (GLP)-1(9–36)amide-mediated cytoprotection is blocked by exendin(9-39) yet does not require the known GLP-1 receptor. *Endocrinology* 2010; 151: 1520–31.

Ban K, Noyan-Ashraf MH, Hoefer J, Bolz SS, Drucker DJ, Husain M: Cardioprotective and vasodilatory actions of glucagon-like peptide 1 receptor are mediated through both glucagon-like peptide 1 receptor-dependent and -independent pathways. *Circulation* 2008, 117:2340-2350.

Bao W, Aravindhan K, Alsaid H, Chendrimada T, Szapacs M, Citerone DR, Harpel MR, Willette RN, Lepore JJ, Jucker BM. Albiglutide, a long lasting glucagon-like peptide-1 analog, protects the rat heart against ischemia/reperfusion injury: evidence for improving cardiac metabolic efficiency. *PLoS One.* 2011;6(8):e23570. Epub. 2011 Aug 26.

Barnett AH (2009) New treatments in type 2 diabetes: a focus on the incretin-based therapies. *Clin. Endocrinol.* (Oxf) 70: 343–353.

Barragan JM, Rodriguez RE, and Blazquez E (1994) Changes in arterial blood pressure and heart rate induced by glucagon-like peptide-1-(7–36) amide in rats. *Am. J. Physiol.* 266:E459–E466.

Barragan JM, Rodriguez RE, Eng J, and Blazquez E (1996) Interactions of exendin-(9–39) with the effects of glucagon-like peptide-1-(7–36) amide and of exendin-4 on arterial blood pressure and heart rate in rats. *Regul. Pept*. 67:63–68.

Basu A, Charkoudian N, Schrage W, Rizza RA, Basu R, Joyner MJ. Beneficial effects of GLP-1 on endothelial function in humans: dampening by glyburide but not by glimepiride. *Am. J. Physiol. Endocrinol. Metab.* 2007; 293: E1289–95.

Bertuglia S and Giusti A. "Microvascular oxygenation, oxidative stress, NO suppression and superoxide dismutase during postischemic reperfusion," *American Journal of Physiology,* vol. 285, no. 3, pp. H1064–H1071, 2003.

Bose AK, Mocanu MM, Carr RD, Brand CL, and Yellon DM. Glucagon-like Peptide 1 Can Directly Protect the Heart Against Ischemia/Reperfusion Injury. *Diabetes* 54:146–151, 2005

Brubaker PL, Drucker DJ. Structure-Function of the Glucagon Receptor Family of G Protein-Coupled Receptors: The Glucagon, GIP, GLP-1, and GLP-2 Receptors. *Receptors and Channels,* 8: 179-188, 2002.

Bullock BP, Heller RS, and Habener JF (1996) Tissue distribution of messenger ribonucleic acid encoding the rat glucagon-like peptide-1 receptor. *Endocrinology* 137:2968–2971.

Buteau J, Roduit R, Susini, S, Prentki M: Glucagon-like peptide-1 promotes DNA synthesis, activates phosphatidylinositol 3-kinase and increases transcription factor pancreatic and duodenal homeobox gene 1 (PDX-1) DNA binding activity in beta (INS-1)-cells. *Diabetologia* 42:856–864, 1999.

Campos RV, Lee YC, Drucker DJ. Divergent tissue-specific and developmental expression of receptors for glucagon and glucagon-like peptide-1 in the mouse. Endocrinology. 1994; 134: 2156-2164. of macromolecules from airways and relaxes pulmonary artery. *Am. J. Physiol. Lung Cell Mol. Physiol.* 1993; 265: L374-L381.

Cerini R, Koshy A, Hadengue A, Lee SS, Garnier P, Lebrec D. Effects of glucagon on systemic and splanchnic circulation in conscious rats with biliary cirrhosis. *J. Hepatol.* 9(1): 69-74, 1989.

Cohn, K.E., Agmon, J., and Gamble, O.W.: Effect of glucagon on arrhythmias due to digitalis toxicity. *Am. J. Cardiol.* 25:683-689, 1970.

D'Alessio DA, Kahn SE, Leusner CR, Ensinck JW. Glucagon-like peptide 1 enhances glucose tolerance by stimulation of insulin release and by increasing insulin-independent glucose disposal. *J. Clin. Invest.* 1994; 93: 2263–2266.

Daniell, H.B., Holl, J.E., Phuett, J.K., Bagwell, E.E., and Woods, E.F.: Cardiovascular effects of glucagon in dogs with nonnodal pacemakers. *J. Electrocardiol.* 3:117-120,1970.

De Witt CR, Waksmann JC: Pharmacology, pathophysiology and management of calcium channel blocker and β-blocker toxicity. *Toxicol. Rev.* 23: 223-238, 2004.

Ding Y, Vaziri ND, Coulson R, Kamanna V and Roh DD. Effects of simulated hyperglycemia, insulin, and glucagon on endothelial nitric oxide synthase expression. Am. J. Physiol. Endocrinol. Metab., 279: E11-E17, 2000.

Ding L, Zhang J. Glucagon-like peptide-1 activates endothelial nitric oxide synthase in human umbilical vein endothelial cells. *Acta. Pharmacologica Sinica* (2012) 33: 75–81.

Drucker DJ, Philippe J, Mojsov S, Chicko WL, Habener JF. Glucagon-like peptide I stimulates insulin gene expression and increases cyclic AMP levels in a rat islet cell line. *Proc. Natl. Acad. Sci. USA*. 1987; 84: 3434-3438.

Drucker DJ. The biology of incretin hormones. *Cell Metab*. 2006; 3: 153-165.

Entman, M.L., Levey, G.S., and Epstein, S.E.: Mechanism of action of epinephrine and glucagon on the canine heart: Evidence for increase in sarcotubular calcium stores mediated by cyclic 3',5'-AMP. *Circ. Res.* 25:429-438, 1969.

Epstein, S.E., Skeleton, C.L., Levey, G.S., and Eentman, M.: Adenyl cyclase and myocardial contractility. *Ann. Intem. Med.* 72:561-578, 1970.

Farah AE. Glucagon and the circulation. Pharmacol Rev 35(3): 181-217, 1983.

Farah, A, and Tuttle, R.: Studies on the pharmacology of glucagon. *J. Pharmacol. Exp. Ther.* 129:49-55, 1960.

Fehmann HC, Göke R, Göke B. Cell and molecular biology of the incretin hormones glucagon-like peptide-I and glucose-dependent insulin releasing polypeptide. *Endocr. Rev.* 1995; 16: 390-410.

Franek E, Gajos G, Gumprecht J, Krętowski A, Zahorska –Markiewicz B, Małecki MT. The role of glucagon -like peptide 1 in glucose homeostasis and in other aspects of human physiology. *Pol. Arch. Med. Wewn.* 2009;119(11):743-51.

Glick G Parmely, WW, Wechsler, AS, and Sonnenblick EH: Glucagon: Its enhancement of cardiac performance in the cat and dog and persistence of its inotropic action despite beta-receptor blockade with propranolol. *Circ. Res. 22*:789-799, 1968.

Golpon HA, Puechner A, Welte T, Wichert PV, Feddersen CO. Vasorelaxant effect of glucagon-like peptide-(7–36)amide and amylin on the pulmonary circulation of the rat. *Regul. Pept.* 2001; 102: 81–6.

Gonzalez-Munoz C, Hernández J. Phosphodiesterases inhibition enhances the effect of glucagon on cardiac automaticity in the isolated right ventricle of the rat. *Physiol. Res.* 2011;60(1):189-92. Epub 2010 Oct 15.

Gonzalez-Muñoz C, Nieto-Cerón S, Cabezas-Herrera J, Hernández-Cascales J. Glucagon increases contractility in ventricle but not in atrium of the rat heart. *Eur. J. Pharmacol.* 2008 Jun 10;587(1-3):243-7. Epub 2008 Apr 8.

Grieve DJ, Cassidy RS, Green BD (2009) Emerging cardiovascular actions of the incretin hormone glucagon-like peptide-1: potential therapeutic benefits beyond glycaemic control? *Br. J. Pharmacol.* 157: 1340–1351.

Hausenloy DJ, Yellon DM: New directions for protecting the heart against ischemia-reperfusion injury: targeting the reperfusion injury salvage kinase (RISK)-pathway. *Cardiovasc. Res.* 61:448–460, 2004.

Holz GG, Heart E, Leech CA: Synchronizing Ca2+ and cAMP oscillations in pancreatic beta-cells: a role for glucose metabolism and GLP-1 receptors? Focus on "regulation of cAMP dynamics by Ca2+ and G protein-coupled receptors in the pancreatic beta-cell: a computational approach". *Am. J. Physiol. Cell Physiol.* 2008, 294:C4-6.

Hough LB, "Genomics meets histamine receptors: new subtypes, new receptors,"*Molecular Pharmacology*, vol. 59, no. 3, pp. 415–419, 2001.

Hui H, Nourparvar A, Zhao X, Perfetti R: Glucagon-like peptide-1 inhibits apoptosis of insulin-secreting cells via a cyclic 5` adenosine monophosphate-dependent protein kinase A- and a phosphatidylinositol 3-kinase-dependent pathway. *Endocrinology* 144:1444–1455, 2003.

Ishibashi Y, Matsui T, Takeuchi M, Yamagishi S. Glucagon-like peptide-1 (GLP-1) inhibits advanced glycation end product (AGE)- induced up-regulation of VCAM-1 mRNA levels in endothelial cells by suppressing AGE receptor (RAGE) expression. *Biochem. Biophys. Res. Commun.* 2010; 391: 1405–8.

Jonassen AK, Brar BK, Mjos OD, Sack MN, Latchman DS, Yellon DM: Insulin administered at reoxygenation exerts a cardioprotective effect in myocytes by a possible anti-apoptotic mechanism. *J. Mol. Cell Cardiol.* 32:757–764, 2000.

Jugdutt BI, "Nitric oxide and cardioprotection during ischemia-reperfusion," *Heart Failure Reviews*, vol. 7, no. 4, pp. 391–405, 2002.

Kaplan YC, Hocaoglu N, Oransay K, Kalkan S, Tuncok Y. Effect of glucagon on amitriptyline-induced cardiovascular toxicity in rats. *Hum. Exp. Toxicol.* 2008 Apr;27(4):321-5.

Kavianipour M, Ehlers MR, Malmberg K, Ronquist G, Ryden L, Wikström G, Gutniak M. Glucagon-like peptide-1 (7-36) amide prevents the accumulation of pyruvate and lactate in the ischemic and non-ischemic porcine myocardium. *Peptides* 2003, 24:569-78.

Komatsu R, Tsushima N, Matsuyama T. Effects of glucagon administration on microcirculation and blood rheology. *Clin. Hemorheol. and Microcirc.* 17(4): 271-277, 1997.

Lipski, J.L., Kaminsky D.M., Donoso, E., and Friedbebg, C.K.: Electrophysiologic properties of glucagon on the normal canine heart. *Am. J. Cardiol.* 23:124-125, 1969.

Lok S, Kuijper JL, Jelinek LJ, Kramer JM, Whitmore TE, Sprecher CA, Mathewes S, Grant FJ, Biggs SH, Rosenberg GB, Sheppard PO, O'Hara PJ, Foster D and Kindsvogel W. The human glucagon receptor encoding gene: structure, cDNA sequence and chromosomal localization. *Gene.* 140: 203-209, 1994.

Love JN, Howell JM. Glucagon therapy in the treatment of symptomatic bradycardia. *Ann. Emerg. Med.* 1997; 29:181-83.

Love JN, Sachdeva DK, Bessman ES, Curtis LA, Howell JM. A potential role for glucagon in the treatment of drug-induced symptomatic bradycardia. *Chest.* 1998 Jul;114(1):323-6.

Love JN, Tandy TK. Beta adrenoreceptor antagonist toxicity: a survey of glucagon availability [letter]. *Ann. Emerg. Med.* 1993; 22:267-68.

Lucchesi, B.R., Stutz, D.R., and Winfield, R.A.: Glucagon: Its enhancement of atrioventricular nodal pacemaker activity and failure to increase ventricular automaticity in dogs. *Circ. Res.* 25:183-190, 1969.

Lucchesi, B.R.: Cardiac actions of glucagon. Circ Res 22:777-787, 1968.

Lund PK. The discovery of glucagon-like peptide 1. *Regul. Pept.* 2005 Jun 15;128(2):93-6.

Madsbad S: Clinical Trial Report-Liraglutide Effect and Action in Diabetes (LEAD™) trial. *Expert Review of Endocrinology and Metabolism* 2009, 4:119-29.

Masini E, Bianchi S, Mugnai L, Gambassi F, Lupini M, Pistelli A, Mannaioni PF. “The effect of nitric oxide generators on ischemia reperfusion injury and histamine release in isolated perfused guinea-pig heart,” *Agents and Actions,* vol. 33, no. 1-2, pp. 53–56, 1991.

Meier JJ, Nauck MA, Kranz D, Holst JJ, Deacon CF, Gaeckler D, Schmidt WE, Gallwitz B. Secretion, degradation, and elimination of glucagon-like peptide 1 and gastric inhibitory polypeptide in patients with chronic renal insufficiency and healthy control subjects. *Diabetes.* 2004; 53: 654-662.

Meneilly GS, McIntosh CH, Pederson RA, Habener JF, Gingerich R, Egan JM, Elahi D: Glucagon-like peptide-1 (7-37) augments insulin-mediated glucose uptake in elderly patients with diabetes. *J. Gerontol. A. Biol. Sci. Med. Sci.* 56:M681–M685, 2001

Mocanu MM, Bell RM, Yellon DM: PI3 kinase and not p44/42 appears to be implicated in the protection conferred by ischemic preconditioning. *J. Mol. Cell Cardiol.* 34:661–668, 2002.

Nathanson D, Erdogdu O, Pernow J, Zhang Q, Nystrom T. Endothelial dysfunction induced by triglycerides is not restored by exenatide in rat conduit arteries *ex vivo. Regul. Pept.* 2009; 157: 8–13.

Nikolaidis LA, Doverspike A, Hentosz T, Zourelias L, Shen YT, Elahi D, and Shannon RP (2005) Glucagon-like peptide-1 limits myocardial stunning following brief coronary occlusion and reperfusion in conscious canines. *J. Pharmacol. Exp. Ther.* 312:303–308.

Nikolaidis LA, Elahi D, Hentosz T, Doverspike A, Huerbin R, Zourelias L, Stolarski C, Shen Y-T, and Shannon RP (2004a) rGLP-1 increased myocardial glucose uptake and improved LV performance in conscious dogs with pacing inducedn dilated cardiomyopathy. *Circulation* 110:955–961.

Nikolaidis LA, Mankad S, Sokos GG, Miske G, Shah A, Elahi D, and Shannon RP (2004b) Effects of glucagon-like peptide-1 in patients with acute myocardial infarction and left ventricular dysfunction after successful reperfusion. *Circulation* 109:962–965.

Nyström T, Gutniak MK, Zhang Q, Zhang F, Holst JJ, Ahrén B, Sjöholm A. Effects of glucagon-like peptide-1 on endothelial function in type 2 diabetes patients with stable coronary artery disease. A*m. J. Physiol. Endocrinol. Metab.* 2004; 287: E1209–15.

Okamura T, Miyazaki M, and Toda N. Responses of isolated dog blood vessels to glucagon. *Eur. J. Pharmacol.* 125: 395-401, 1986.

Ørskov C, Wettergren A, Holst JJ. Biological effects and metabolic rates of glucagonlike peptide-1 (7-36) amide and glucagon like peptide-1 (7-37) in healthy subjects are indistinguishable. *Diabetes.* 1993; 42: 658-661.

Persson K, Gingerich RL, Nayak S, Wada K, Wada E, Ahrén B. Reduced GLP-1 and insulin responses and glucose intolerance after gastric glucose in GRP receptor-deleted mice. *Am. J. Physiol. Endocrinol. Metab.* 2000; 279: E956-E962.

Pohl SL, Birnbaumer L and Rodbell M. Glucagon-sensitive adenyl cylase in plasma membrane of hepatic parenchymal cells. *Science* 164: 566-567, 1969.

Pruett JK, Woods EF, Daniell HB. Glucagon enhanced automaticity in spontaneously beating Purkinje fibres of canine false tendons. *Cardiovasc Res.* 1971 Oct;5(4):436-9.

Redondo A, Trigo MV, Acitores A, Valverde I, Villanueva-Penacarrillo ML: Cell signalling of the GLP-1 action in rat liver. *Mol. Cell Endocrinol.* 204:43–50, 2003.

Richter G, Feddersen O, Wagner U, Barth P, Göke R, Göke B. GLP-1 stimulates secretion of macromolecules from airways and relaxes pulmonary artery. *Am. J. Physiol. Lung. Cell Mol. Physiol.* 1993; 265: L374-L381.

Ritzel R, Orskov C, Holst JJ, Nauck MA. Pharmacokinetic, insulinotropic and glucagonostatic properties of GLP-1 [7–36] amide after subcutaneous injection in healthy volunteers. *Diabetologia.* 1995; 38: 720–725.

Rosic M, Andjelkovic I, and Zlokovic B. "Characterization of 3H histamine transport at sarcolemal membrane of the isolated perfused guinea-pig heart in the presence of glucagons and H1 and H2 receptor antagonists," *Biomedical and Biochemical Acta,* vol. 46, no. 10, pp. 736–740, 1987.

Rosic M, Andjelkovic I, Zlokovic B et al., "Effects of glucagon and H1 and H2 receptor antagonists on 3H histamine transport at sarcolemmal membrane of the isolated perfused guinea pig heart," Yugoslav *Physiological and Pharmacological Acta*, vol. 24, supplement, p. 391, 1988.

Rosic M, Pantovic S, Rosic G, Tomic-Lucic A, Labudovic T, Zivkovic V, Jakovljevic V. Glucagon effects on ischemic vasodilatation in the isolated rat heart. *J Biomed Biotechnol*. 2010;2010:231832. Epub 2010 Mar 18.

Ross G. Regional circulatory effects of pancreatic glucagon. *Br. J. Pharmac.* 38:735-742, 1970.

Rouillé Y, Martin S, Steiner DF. Differential processing of proglucagon by the subtilisin-like prohormone convertases PC2 and PC3 to generate either glucagon or glucagon-like peptide. *J. Biol. Chem.* 1995; 270: 26 488-26 496.

Runge S, Thogersen H, Madsen K, Lau J, Rudolph R: Crystal structure of the ligand-bound glucagon-like peptide-1 receptor extracellular domain. *J. Biol. Chem.* 2008, 283:11340-7.

St Onge EL, Miller SA (2010) Albiglutide: a new GLP-1 analog for the treatment of type 2 diabetes. *Expert. Opin. Biol. Ther.* 10: 801–806.

Steiner, C, Wrr, A.L., and Damato, A.M.: Effects of glucagon on atrioventricular conduction and ventricular automaticity in dogs. *Circ. Res.* 24:167-177, 1969.

Stuhlinger W, Turnheim K, Gmeiner R. The effects of glucagon on the pulmonary circulation in the dog. *European Journal of Pharmacology* 28(1): 241-243, 1974.

Suda K, Takahashi H, Fukase N, et al. Distribution and molecular forms of glucagon-like peptide in the dog. *Life Sci.* 1989; 45: 1793-1798.

Tolins JP. Mechanisms of glucagon-induced renal vasodilation: role of prostaglandins and endothelium-derived relaxing factor. *J. Lab. Clin. Med.* 120: 941-948, 1992.

Tsao PS, Aoki N, Lefer DJ, Johnson G III and Lefer A, "Time course of endothelial dysfunction andmyocardial injury duringmyocardial ischemia and reperfusion in the cat," *Circulation,* vol. 82, no. 4, pp. 1402–1412, 1990.

Uesaka T, Yano K, Sugimoto S, Ando M: Glucagon-like peptide isolated from the eel intestine: effects on atrial beating. *J. Exp. Biol.* 2001, 204:3019-26.

Varro, V. and Czernay, L. (1966). The effect of intra-arterial insulin and glucagon on the glucose metabolism of the small intestine in the dog. *Scand. J. Gastroenterol.,* 1, 232-237.

Vila Petroff MG, Egan JM, Wang X, Sollott SJ: Glucagon-like peptide-1 increases cAMP but fails to augment contraction in adult rat cardiac myocytes. *Circ. Res.* 2001, 89:445-52.

Wakelam MJ, Murphy GJ, Hruby VJ and Houslay MD. Activation of two signal-transduction systems in hepatocytes by glucagon. *Nature* 323: 68-71, 1986.

Wei Y and Mojsov S (1996) Distribution of GLP-1 and PACAP receptors in human tissue. *Acta. Physiol. Scand.* 157:355–357.

White CM: A review of potential cardiovascular uses of intravenous glucagon administration. *J. Clin. Pharmacol.* 39: 442-447, 1999.

Whitehouse FW, and James TN: Chronotropic action of glucagon on the sinus node. *Proc. Soc. Exp. Biol. Med.* 122:823-826, 1966.

Wilkerson RD, Pruett JK, Woods EF. Glucagon-enhanced ventricular automaticity in dogs. Its concealment by positive chronotropism. *Circ. Res.* 1971 Dec;29(6):616-25.

Williams JF Jr, Childress RH, Chip JN and Border JF. Hemodynamic effects of glucagon in patients with heart disease. *Circulation* 29: 38-47, 1969.

Wilmen A, Walkenbach A, Füller P, Lankat-Buttgereit B, Göke R, Göke B. The genomic organization of the human GLP-1 receptor gene. *Exp. Clin. Endocrinol. Diabetes.* 1998; 106: 299-302.

Xiao Yong-Fu, Nikolskaya Alena, Jaye Deborah A, Sigg Daniel C. Glucagon-like peptide-1 enhances cardiac L-type Ca^{2+} currents via activation of the cAMP-dependent protein kinase A pathway. *Cardiovascular Diabetology* 2011, 10:6.

Yamamoto H, Lee CE, Marcus JN, Williams TD, Overton JM, Lopez ME, Hollenberg AN, Baggio L, Saper CB, Drucker DJ, et al. (2002) Glucagon-like peptide-1 receptor stimulation increases blood pressure and heart rate and activates autonomic regulatory neurons. *J. Clin. Investig*. 110:43–52.

Yu M, Moreno C, Hoagland KM, Dahly A, Ditter K, Mistry M, Roman RJ. Antihypertensive effect of glucagon-like peptide 1 in Dahl saltsensitive rats. *J. Hypertens* 2003; 21: 1125–35.

Zhao T, Parikh P, Bhashyam S, Bolukoglu H, Poornima I, Shen YT, et al: Direct effects of glucagon-like peptide-1 on myocardial contractility and glucose uptake in normal and postischemic isolated rat hearts. *J. Pharmacol. Exp. Ther.* 2006, 317:1106-13.

In: Glucagon
Editor: Ahmed Al-Jebawi

ISBN: 978-1-62618-147-2

Chapter 4

The Role of Abnormal Glucagon in Type 2 Diabetes

Ahmed F. Al-Jebawi*

Indiana University School of Medicine, Indianapolis, Indiana, US
Dubai Diabetes Center, Dubai, United Arab Emirates

Abstract

In addition to the abnormal glucagon response to arginine, both fasting and postprandial hyperglucagonemia were demonstrated in patients with type 2 diabetes. Furthermore, patients with type 2 diabetes may suffer from an abnormal response of glucagon to hypoglycemia. The most likely reason for this poor response is the reduction of intraislet insulin in response to falling blood glucose. These findings had recently contributed to the development of new agents that target glucagon among patients suffering from diabetes mellitus type 2. The inhibitory effect of GLP-1 on glucagon appears to play an important role for the antidiabetic actions of the incretin based therapeutic agents, including the dipeptidyl peptidase-4 (DPP-4) inhibitors and the GLP-1 receptor agonists. Although Somatostatin analogs were demonstrated to reduce glucagon secretion, such agents may also inhibit the release of other hormones

* Correspondence to: Ahmed F. Al-Jebawi, MD FACP FACE, E-mail: aljebawi@hotmail.com, Clinical Assistant Professor, Indiana University School of Medicine, Indianapolis, Indiana USA, Chairman, Diabetes Diploma Program, Dubai Diabetes Center, Dubai, United Arab Emirates.

including Insulin. This undesired effect of somatostatin on other hormones involved in glucose homeostasis made many investigators search for more selective somatostatin analogs. Glucagon antibodies, glucagon receptor gene inactivating agents, glucagon receptor signaling inhibitors and glucagon receptor antagonists are still being evaluated for their therapeutic potential.

Introduction

As our understanding of the mechanism of diabetes type 2 has progressed so significantly over the last 2 decades, there has been greater interest in developing new therapeutic strategies to improve the effectiveness of diabetes treatment. Although our understanding of the contribution of different hormones to the occurrence of hyperglycemia had created many new opportunities for clinical work, it made us realize that the pathogenesis of this disease is much more complex than what we thought in the past. In fact, the complex interactions among glucose regulatory hormones may result in a self-perpetuating process that may make it quite difficult to establish a cause-effect relationship.

The notion that diabetes mellitus might, at least in part, be a consequence of excessive glucagon secretion was first reported by Rodriquez-Candela in 1947. In his report, Rodriquez-Candela showed that diabetes of dogs with ligated pancreatic ducts could be ameliorated by removal of the pancreatic remnant, suggesting that the presence of alpha cells contributed to the severity of the diabetes [1]. The role of glucagon in the pathogenesis of diabetes was confirmed in 1969 when Roger Assan and co-workers showed that glucagon levels are very high among patients suffering from diabetic ketoacidosis [2]. In 1970, Unger et al. proved that diabetic subjects' alpha cells are producing excessive amounts of glucagon. Although the mean fasting level of glucagon among type 2 diabetic subjects did not differ significantly from the non-diabetic subjects, his study revealed that in non-diabetic subjects glucagon secretion is diminished by induced hyperglycemia to levels significantly below that of diabetics with comparable degrees of spontaneous hyperglycemia. Unger et al. stated: "It is reasonable to suppose that the inappropriately high glucagon, as well as the inappropriately low insulin levels, relative to the prevailing glucose level, are influencing the net hepatic glucose balance and increasing the severity of hyperglycemia, and the amount of insulin required to reduce hepatic glucose production to normal" [3]. In 1975, Gerich and his colleagues found that, in people with type 1 diabetes, infusion of somatostatin delayed the development of hyperglycemia and ketosis after withdrawal of insulin therapy [4]. Furthermore, they found out that somatostatin lowered plasma glucose concentrations in insulin-withdrawn, hyperglycemic patients [5]. In the absence of insulin secretion among those patients with type 1 diabetes, the somatostatin action on glucose was presumed to be through its suppressive effects on glucagon secretion. Despite the emergence of this significant bihormonal hypothesis to explain the pathophysiology of diabetes, alpha cells and glucagon were overlooked for more than 40 years.

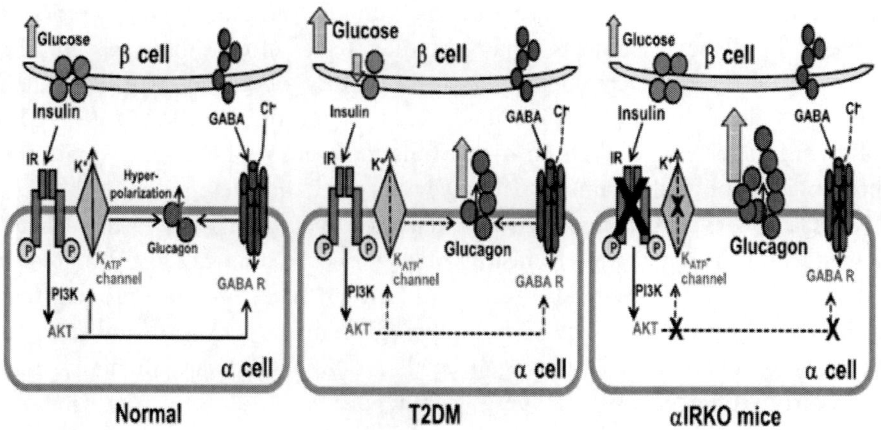

Figure 1. Regulation of glucagon by insulin in pancreatic alpha cells. In normal alpha cells, glucagon secretion is controlled by autocrine and paracrine processes (left). In addition, glucagon secretion is regulated by intrinsic alpha cell mechanisms (not shown). Increased glucose levels cause insulin release from beta cells. Through binding to alpha cell receptors (IR), insulin activates the PI3K/AKT pathway leading to either opening the KATP channels or the recruitment of GABA-activated GABA receptor Cl_ channels (GABA R; channels that are activated by GABA released from neighboring beta cells). Both processes culminate in membrane hyperpolarization and inhibition of glucagon release. In patients with T2DM (middle), reduced insulin secretion leads to less activation of alpha cell insulin receptors and downstream processes will be correspondingly reduced, causing oversecretion of glucagon. In aIRKO islets (right) insulin signaling is abolished, and the relief from the paracrine suppression normally exerted via the insulin receptors account for stimulation of glucagon secretion. Cell Metabolism. Apr; 9(4):303-5 *(with permission)*.

Characteristic Abnormalities of Glucagon Secretion in Type 2 Diabetes

Much work has been done on understanding how pancreatic alpha cell function and proliferation are regulated among normal and diabetic people. The finding that reduction of glucagon in pancreatectomized canines caused a marked decrease in hepatic glucose production [6] and suppression of glucagon in diabetic humans improved glucose tolerance [7, 8] had opened up new opportunities for targeting hyperglycemia among diabetic people. Although many experts believe that the number of glucagon receptors is reduced in diabetic subjects, many studies have shown that, despite the presence of fewer glucagon receptors, the ability of glucagon to stimulate cAMP production may remain unchanged or even be elevated [9]. The following underlying defects in α-cell function were proposed to explain the metabolic abnormalities among patients with type 2 diabetes:

Fasting Hyperglucagonaemia: Reaven et al. had shown that many diabetic subjects have elevated fasting glucagon levels that can be 50% greater than non-diabetic individuals [10, 11]. Further investigations proved that despite the suppressive effects of both hyperglycemia and hyperinsulinemia on glucagon production, basal glucagon levels are higher in many diabetics than in control subjects [10]. Hyperglycemia produces an apparent glucose unresponsiveness of alpha cells probably by down-regulating the glucose sensing sites that mediate the negative glucagon secretory response to a rise in glucose. It is also probable that these cells continue to produce large amounts of glucagon as a consequence of insulin deficiency and/ or insulin resistance. Li and co-workers had shown that short term glucagon infusion continues to stimulate glucose production in the long term. They reported that glucagon infusion for 4 weeks caused persistent hyperglycemia in mice. Interestingly, those effects of glucagon were prevented by co-infusion of the glucagon antagonist [Des-His1-Glu9] glucagon [12, 13]. Thus the results document a sustained glycemic effect of hyperglucagonaemia. As a matter of fact, the glucagon-infused animals had developed features of diabetic nephropathy with increased kidney weight/body weight ratios, albuminuria and glomerular mesangial expansion [12]. Probably, these effects are caused by hyperglycemia rather than the high levels of glucagon itself. Since basal glucagon is responsible for sustaining two-thirds of basal hepatic glucose output in normal humans, Baron et al. investigated its role in sustaining elevated rates of basal hepatic glucose output in diabetes type 2. They demonstrated that basal hepatic glucose output is higher in diabetics than in controls [10]. Since fasting hyperglucagonemia was also described in persons with impaired glucose tolerance [14-16], it was suggested that there is an early phase of involvement in both α- and β-cell dysfunction in pre-diabetic states.

Postprandial Hyperglucagonaemia: In addition to the fasting hyperglucagonaemia, type 2 diabetes is characterized by a significant lack of glucagon suppression during periods of postprandial hyperglycemia. Mitrakou et al. had shown that both impaired early insulin release and excessive glucagon secretion contribute to a significant reduction in postprandial hepatic glucose uptake in patients with type 2 diabetes [17]. Recent studies performed by Woerle et al. indicated that, in type 2diabetes, the release of ingested and endogenously produced glucose into the systemic circulation are both increased postprandially. The postprandial increased endogenous glucose release involves both increased gluconeogenesis and glycogenolysis [18]. The abnormal glucagon response to oral glucose load was confirmed in many clinical studies [11]. What is more interesting is the observation that plasma glucagon level paradoxically rose after glucose administration to a group with severe diabetes (whose fasting blood glucose exceeded 200 mg/100 ml) [19].

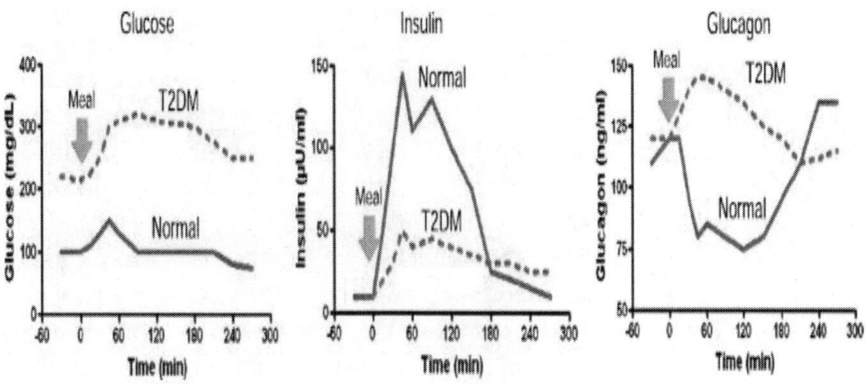

Figure 2. Plasma insulin, glucagon, and glucose levels in response to a meal in healthy individuals and patients with T2DM. Blood glucose levels are regulated by carefully orchestrated changes of insulin and glucagon secretion in response to carbohydrate meal ingestion in healthy individuals (Normal). An increased plasma glucose (left) following a meal evokes insulin secretion from the beta cells (middle); insulin secretion in individuals with T2DM is reduced (red-dotted line). Glucagon secretion is normally suppressed by the elevation of plasma glucose (right; blue line). In diabetic patients, the suppressor effect of glucose is reduced, and it may even stimulate glucagon release (right; red line). Cell Metabolism. Apr; 9(4):303-5 *(with permission).*

Hyper-Responsiveness to Arginine

Amino acids are considered as potent stimuli for glucagon secretion. Thus, ingestion of protein meals [20] or stimulation with IV arginine [21] results in greater relative stimulation of plasma glucagon in T2DM subjects compared with nondiabetic controls. Arginine provocative test is performed to stimulate the secretion of glucagon from the alpha cells. Arginine-stimulated hyperglucagonemia in patients with diabetes was first discovered by Aronoff et al. [22, 23]. They infused arginine to 3 groups of 27 nondiabetic Pima Indians, 34 nondiabetic Caucasians, and 12 diabetic Pima Indians with recent onset of their disease. Although the fasting glucagon levels in the three groups were not significantly different, the diabetic Pimas had glucagon levels significantly higher than the normo-glycemic Pimas during the arginine infusion. Plasma insulin levels also increased during the infusion but, notably, never differed significantly between these two groups. There was no significant difference in the glucagon levels at any sampling point between the nondiabetic Pimas and Caucasians [22]. Their results demonstrated that hypeglucagonaemia or an elevated ratio of glucagon to insulin play a major role in the development of endogenous hyperglycemia in diabetes through increasing the hepatic glucose production. Among patients with type 1 diabetes, an absolute deficiency in insulin secretion was thought to allow the arginine to cause an exaggerated response of alpha cells to secrete glucagon. While insulin treatment was found to correct this phenomenon among patients with type 1 diabetes [21, 25], those patients with type 2 diabetes continue to demonstrate hyper-responsiveness to Arginine [24]. Tsuchiyama et al. had clearly shown that, after arginine challenge, glucagon levels in patients with diabetes type 1are negatively correlated with C-peptide levels. In contrast, glucagon levels in diabetes type 2 are positively correlated with C-peptide levels [26]. These results suggest that the mechanisms involved in the glucagon response to arginine challenge are distinctly different between type1 and type 2 diabetic patients. In patients with type 2 diabetes, the exaggerated glucagon response to arginine is most probably related to alpha cell insulin resistance. While some studies have shown that exogenous insulin replacement can reduce the glucagon levels to the normal range [27], other studies did not observe any reduction in glucagon levels among type 2 diabetic patients with severe hyperinsulinemia [24, 26]. If the response of postprandial hyperglucagonemia to exogenous insulin is proved to be impaired, treatment with glucagon lowering agents and/ or glucagon antagonists in combination with insulin therapy will be clinically justified.

Abnormal Response to Hypoglycemia

Glucagon provides the body's primary defense against hypoglycemia. Hypoglycemia is associated with a decrease in insulin signaling and a glucagon mediated increase in the activation of AMP-activated protein kinase (AMPK). These changes were in turn associated with an augmentation of glucagon's stimulatory effect on glycogen phosphorylase activity and its inhibitory effects on glycogen synthase activity. The disruption in insulin signaling thereby allows glucagon's effect on cAMP-activated protein kinase (PKA) and AMPK to drive glycogen breakdown in a life-saving manner. PKA has long been established as the major intracellular mediator of glucagon's glucoregulatory effects within the hepatocyte, and prominently coordinates the phosphorylation of glycogen synthase and glycogen phosphorylase, resulting in net glycogen breakdown [28]. Although there are different possible mechanisms for the rise in glucagon level during periods of hypoglycemia, the most acceptable model is thought to be through the reductions of intraislet insulin in response to falling blood glucose. When intraislet insulin drops, the α-cells release more glucagon. To maximize the stimulation of hepatic glucose production, glucagon is usually secreted in a pulsatile fashion [29]. One of the most difficult problems to treat among people with diabetes is when their alpha cells fail to respond appropriately to hypoglycemia. Although this problem usually represents a major challenge in type 1 patients, those with type 2 may also have variable degrees of reduced glucagon responses to hypoglycemia [30]. As the brain is not capable of synthesizing glucose and storing adequate amounts of glycogen, it totally depends on the continuous supply of glucose from the circulation. Therefore, any defect in counter-regulatory hormones can significantly increase the risk of developing the serious complications of severe hypoglycemia. In addition to these defects in glucagon production, diabetic patients may suffer from compromised adrenomedullary and sympathetic neural responses to hypoglycemia [31]. In the presence of normal glucagon release, plasma glucose recovery from Insulin-induced hypoglycemia is normal despite epinephrine deficiency. In the absence of glucagon release, epinephrine release is augmented and partially compensates so that recovery from hypoglycemia, although somewhat attenuated, still occurs. Only in the absence of both glucagon and epinephrine release does recovery from hypoglycemia fail to occur [25]. Therefore, when plasma glucose concentrations fall, the first physiological defense is an immediate drop in beta cell insulin secretion. This step would also signal an increase in alpha cell glucagon secretion. When alpha cells fail to increase the amount of glucagon secretion, the adrenomedullary epinephrine secretion would increase [31-33]. As patients with type 1 diabetes and advanced type 2 diabetes do not have a normal glucagon response to hypoglycemia and are at risk for hypoglycemia-associated autonomic failure, they frequently suffer from vicious cycle of recurrent iatrogenic hypoglycemia [34]. The risk of severe iatrogenic hypoglycemia among these patients can reach at least 25-folds compared with those with an intact epinephrine response [35]. Furthermore, failure of glucagon and epinephrine secretion may lead to loss of the warning symptoms of hypoglycemia.

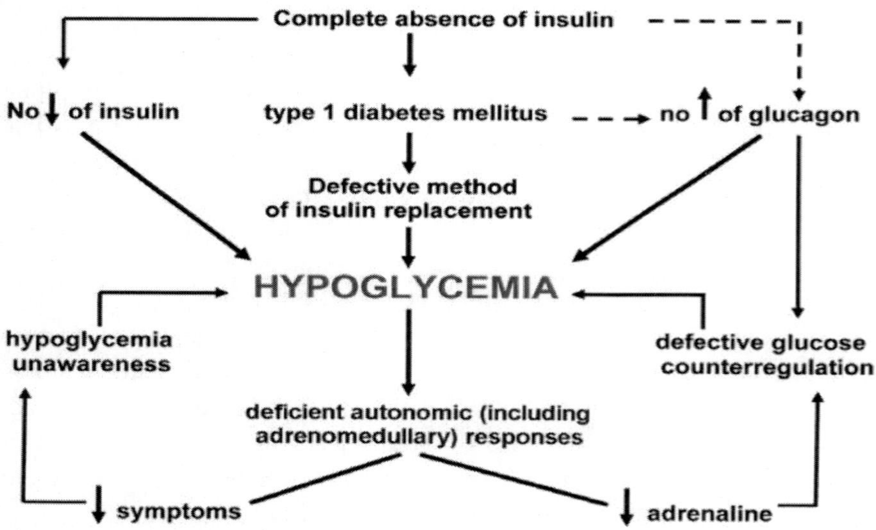

Figure 3. Hypoglycemia-Associated Autonomic Failure. The vicious circle of deficiency of autonomic nervous system due to hypoglycemia. Diabetes Research and Clinical Practice 93 S (2011) S 92–S 96 (with permission).

Reduced ability or failure to recognize hypoglycemia at the physiological plasma glucose concentration at which warning symptoms normally occur is usually referred to as "hypoglycemia unawareness". Patients unaware of hypoglycemia either do not feel symptoms of hypoglycemia or ultimately feel the symptoms, but at much lower plasma glucose levels (high thresholds). As a matter of fact, the definition of hypoglycemia may get revised based on the amount of glucagon and the other counter-regulatory hormones normally secreted in response to decrements in glucose. Traditionally, hypoglycemia is defined as plasma glucose concentration below 50 mg/dl (2.7 mmol/l) [36, 37]. However, the fact that the counter-regulatory hormones are normally released in response to hypoglycemia at higher plasma glucose concentrations i.e. at a plasma glucose concentration of 65-68 mg/dl (3.5 mmol/l) would argue that the safe glycemic targets for intensive therapy of diabetes should be defined according to these physiological responses [figure 3]. As the risk of suffering from severe hypoglycemic attacks is much higher among those with unawareness [38, 39], such patients should not be subjected to intensive therapy with Insulin or Insulin secretagogues.

When people with diabetes are treated with insulin or insulin secretagogues such as sulfonylureas or glinides, their glucose levels may drop to critically low values subjecting them to different complications or even death. Although this problem represents a real threat to diabetic patients, the fear of hypoglycemia among patients and their health care providers may frequently be overestimated leading to under-treatment of hyperglycemia and consequently failure to achieve a good glycemic control.

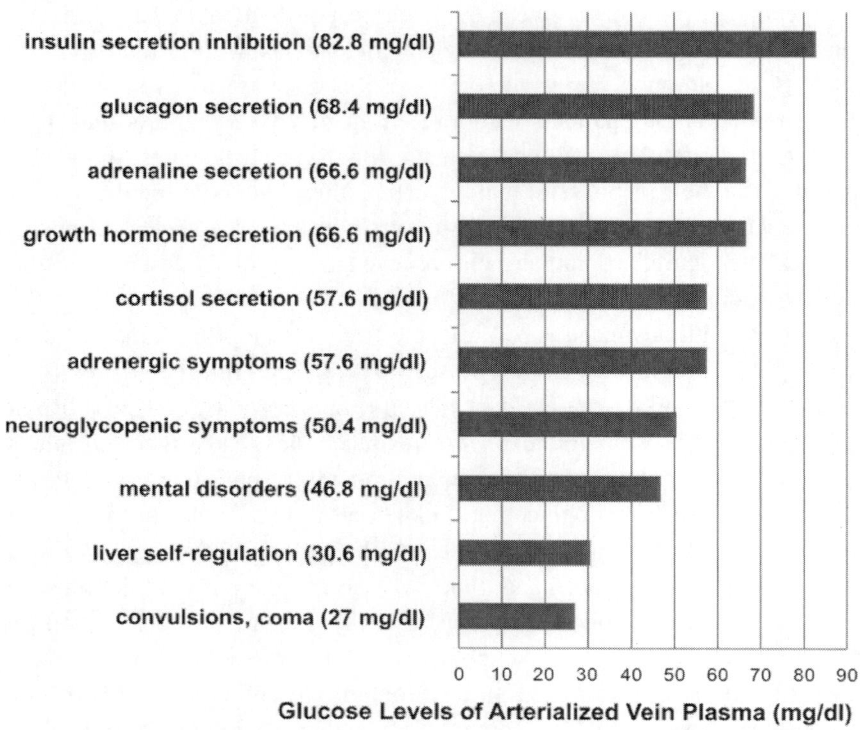

Figure 4. Strict chain of the glycemic threshold of glucose counterregulation. Diabetes Research and Clinical Practice 93 S (2011) S 92–S 96 (with permission).

Pathogenesis of Hypoglycemia Unawareness

As indicated in figure 3, the glucose counter-regulation in normal people follows a specific chain of reactions. Obviously, patients who develop hypoglycemia due to Insulin treatment are not able to have any reduction in their Insulin levels as they continue to absorb the injected Insulin despite their low glucose levels. Moreover, glucagon cannot increase in many patients with type 1 diabetes or advanced type 2diabetes [40, 41]. Although the deficient glucagon response most likely occurs as a result of deficiency of endogenous insulin secretion [42, 43], the high therapeutic level of Insulin is a major contributing factor in glucagon suppression. Moreover, the adrenaline response to the constant decrease in glucose level is often impaired [44, 45]. When hypoglycemic episodes occur frequently, they usually result in moving the glycemic threshold for the adrenaline response to a lower level. This phenomenon is particularly observed among patients with a tight glycemic control [45]. Although the protective glucagon response to hypoglycemia is usually normal among patients with a recent history (1-2 years) of diabetes, it begins to fail afterwards. Most patients with type 1 diabetes show some impairment in their response to hypoglycemia after 2 years of the diagnosis of diabetes. After five years, almost all subjects with type 1 diabetes suffer from a remarkable degree of impairment or even a complete absence of their protective glucagon response [40].

Subjects with advanced type 2 diabetes may develop the same metabolic defects [34]. The mechanism(s) of the loss of the glucagon response to hypoglycemia is unknown. The fact that glucagon responses to stimuli other than hypoglycemia were found to remain intact [46], had suggested that an abnormality of signaling to alpha cells is probably responsible for the loss of glucagon response to hypoglycemia. Fukuda et al. had found that loss of the glucagon response is strongly associated with the loss of endogenous insulin secretion [47]. On the other hand, there was no association with the presence of classical diabetic autonomic neuropathy [48]. Banarer S et al. proved that the glucagon response to hypoglycemia is prevented by intraislet hyperinsulinemia despite an intact autonomic—adrenomedullary (plasma epinephrine), sympathetic neural (plasma norepinephrine), and parasympathetic neural (plasma pancreatic polypeptide) response and a low alpha cell glucose concentration in humans [42]. During tolbutamide infusion, they showed that glucagon response was markedly reduced while most of the other responses to hypoglycemia remained intact.

Targeting Glucagon Action as a Therapeutic Strategy

Treatment of diabetes has traditionally focused on replacing Insulin, stimulating its endogenous production and/ or enhancing its effectiveness. The recent understanding of glucagon abnormalities among diabetic people had opened a new field of treatment. Since hyperglucagonemia significantly contributes to hyperglycemia, suppressing glucagon action has been an attractive approach to reverse the metabolic disorders of diabetes. As glucagon antagonism may lead to compensatory glucagon hypersecretion and alpha cell hyperplasia, the efficacy of such treatment on lowering the glucose level may potentially get compromised. Furthermore, the very high glucagon levels may raise concerns regarding potential development of neoplastic pancreatic lesions. Due to the fact that glucagon plays a primary role in counterregulatory mechanisms during hypoglycemia, the use of glucagon antagonists should be carefully evaluated in patients with T2DM treated with anti-hyperglycemic agents that may lead to the development of hypoglycemia [49].

I.Therapeutic Effects of Somatostatin and Analogs on Glucagon Secretion

The effect of reducing the hyperglycemic consequences of glucagon was first shown when somatostatin administration was able to restore glucose levels to the normal range in insulin-deficient humans and dogs [50-53]. A variety of other methods to block glucagon secretion [54. 55] or action [56, 57] were subsequently used to evaluate the role of glucagon in the pathogenesis of diabetes. Somatostatin inhibits glucagon and insulin release in endocrine islets by interacting with membrane somatostatin receptors [58]. Five distinct membrane receptors (SSTR1–5) for somatostatin are known, and at least two (SSTR2 and SSTR5) have been proposed to regulate pancreatic endocrine function [59]. The fact that somatostatin affects other hormones involved in glucose homeostasis made many investigators search for more selective methods to suppress the glucagon. A recent immunohistochemical study in the rat demonstrated that SSTR2 is expressed at high levels on alpha cells, whereas it is barely detectable on beta cells [60] suggesting a selective role for this receptor in regulating glucagon release. Rossowski et al. showed that SSTR2 selective peptidergic analogs played a major role in inhibiting the glucagon release from pancreatic islets of the rat. [61]. However, data from human pancreatic islets suggest that species differences for SSTR expression may exist [62]. Therefore, further studies are required to clarify the role of the individual SSTRs in regulating hormonal secretion of the endocrine islets.

II.Therapeutic Effects of Incretin Hormones on Glucagon Secretion

Gastrointestinal factors that are clearly stimulated by oral glucose include GLP-1, GLP-2 and GIP. GLP-1 is a powerful inhibitor of glucagon secretion [63] (see chapter 5). GLP-2 and GIP have been found to stimulate glucagon secretion in both physiological and 'pharmacological' studies [64-66]. Although the mechanism of GLP-1-induced inhibition of glucagon is still controversial, a paracrine mechanism involving somatostatin secretion from neighboring δ-cells has experimental support [66]. The inhibitory effect of GLP-1 on glucagon appears to play an important role for the antidiabetic actions of the incretin based therapeutic agents, including the dipeptidyl peptidase-4 (DPP-4) inhibitors and the GLP-1 receptor agonists (for more details, refer to chapter 5).

III.Therapeutic Effects of Immunoneutralization of Endogenous Glucagon

The administration of glucagon antibodies has been used as a model of selective glucagon deficiency in animal diabetes. Flatt and co-workers [56], demonstrated that treatment of hyperglycaemic ob/ob mice with glucagon antibodies caused a normalization of blood glucose (after 6 h) paralleled by a 15-20 % decrease in plasma insulin. Brand et al. produced an extensive immunoneutralization of endogenous glucagon in diabetic rats using excess amounts of a high-affinity, glucagon-free monoclonal glucagon antibody in order to obtain reliable data concerning the effect of selective glucagon deficiency on blood glucose regulation. Monoclonal glucagon antibody therapy was found to normalize blood glucose levels in moderately diabetic rats injected with streptozotocin as neonates. Severely diabetic streptozotocin injected rats are insensitive to glucagon neutralization unless the diabetes is moderately controlled by insulin during or prior to the experiment. The findings of Brand et al. indicated that glucagon antagonism could provide a therapeutic approach to the treatment of hyperglycaemia in moderate type 2diabetes [67].

IV.Therapeutic Effects on Glucagon Receptor

Glucagon acts through a 7-transmembrane class II receptor coupled to G proteins. Glucagon receptor has been cloned from both rodent and human tissue [68, 69]. Receptor activation stimulates adenylate cyclase, resulting in cyclic adenosine monophosphate (cAMP) accumulation and increased protein kinase A activity. The glucagon receptor has also been reported to be independently coupled to a Ca2+-mediated signaling pathway. Glucagon receptor activation leads to an increase in intracellular Ca2+ through activation of phospholipase C [70]. In addition, glucagon activates multiple enzymes required for gluconeogenesis, especially the enzyme system for converting pyruvate to phosphoenolpyruvate, the rate-limiting step in gluconeogenesis [71, 72]. In the hepatocytes, receptor activation leads to increased hepatic glucose production by enhancing both glycogenolysis and gluconeogenesis along with inhibition of glycogenesis [70, 73]. The glucagon receptor is also expressed in extrahepatic tissues, including the heart, intestinal smooth muscle, kidney and brain [74, 75].

A. Effects of Disruption of Glucagon Receptor Gene

To examine the role of glucagon receptor function in glucose homeostasis, Parker et al. generated mice in which the glucagon receptor gene was inactivated (GR^{-} /$^{-}$ mice). Blood glucose levels were found to be somewhat reduced in GR^{-} /$^{-}$ mice relative to wild type, in both the fed and fasted state. Glucose tolerance, as assessed by an oral glucose tolerance test, improved. Plasma glucagon levels were strikingly elevated in both fed and fasted animals. Despite a total absence of glucagon receptors, these animals maintained near-normal glycemia and normal lipidemia, in the presence of circulating glucagon concentrations that were elevated by two orders of magnitude. Interestingly, fasting during 12–16 h did not lead to hypoglycemia [76, 77]. However, fasting for 24 hours did result in lower glucose levels than in wild-type animals (1.7 vs. 3.7 mM) [77]. Mice with a null mutation in the glucagon receptor gene manifest lower blood glucose levels as compared with wild-type mice and display alpha cell hyperplasia [76, 77].

Human mutations in the glucagon receptor gene were recently described [78]. The discovery of such mutations came after the attempts to explain the severe hyperglucagonemia in patients presenting with a pancreatic mass (or pancreatic alpha cell hyperplasia) without glucagonoma syndrome. Although mild hyperglucagonemia can be encountered in a variety of clinical conditions such as hypoglycemia, pancreatitis, cirrhosis, renal failure, and acute trauma or burns, significant hyperglucagonemia is usually caused by glucagonoma. In the absence of glucagonoma syndrome, such glucagon receptor mutations should be highly suspected [78]. Zhou et al. found a homozygous missense mutation (P86S) in the glucagon receptor gene resulting in an ineffective glucagon receptor. The inactivating mutation explains the absence of glucagonoma syndrome in these patients, as it renders the glucagon receptor much less responsive to glucagon. Few other mutations were described [79].

B. Effects of Inhibition of Glucagon Receptor Signaling

Knockdown of glucagon receptor expression in mice using antisense oligonucleotides lead to improved glucose metabolism in *ob/ob* mice and decreased hepatic glucose production [80, 81]. Using antisense oligonucleotides (ASOs) to reduce target gene expression is a novel approach for treating various diseases, including metabolic disorders [82, 83]. GR-ASO treatment decreased glucagon receptor mRNA expression in liver by 83%, reduced glucagon-stimulated hepatocyte cAMP formation, significantly ameliorated the diabetic syndrome, markedly improved the glucose handling during an oral glucose tolerance test, and lowered hyperglycemic response to a glucagon challenge [81]. Liang et al. showed that suppressive effect of GR-ASO treatment on glucagon receptor expression, primarily in liver, is sufficient to reduce blood levels of glucose, triglycerides, and fatty acids in diabetic animals. In contrast to glucagon receptor knockout mice, there was an increase in plasma insulin levels in GR-ASO–treated *db/db* mice. This effect might be the result of an increase in glucagon like peptide 1 (GLP-1) levels in the pancreas as a result of increased glucagon levels observed in GR-ASO–treated mice. A small amount of proglucagon is normally processed to GLP-1 in pancreatic islet cells [77]. In glucagon receptor knockout mice, pancreatic alpha-cell hyperplasia increases the content of total GLP-1 in pancreatic extracts by 25-fold [77]. Elevated islet GLP-1 levels may act in a paracrine manner to stimulate insulin release. As a consequence, blood glucose control might benefit from an increase in circulating insulin levels, which may also result in a reduction of free fatty acids and triglyceride levels, as was observed in GR-ASO–treated *db/db* mice. The use of GR-ASO was associated with elevated liver triglyceride levels, which is similar to the observations made in liver-specific phosphoenolpyruvate carboxykinase (PEPCK) knockout mice [84]. In these knockout mice, hepatic gluconeogenesis was severely interrupted because of the lack of PEPCK. Considering that glucagon plays an important role in regulating PEPCK expression, reduced glucagon receptor levels could reduce PEPCK activity in hepatocytes and thus somewhat mimic the consequence of PEPCK knockout mice. Another possible explanation of increased triglyceride content in liver could be related to the effect of glucagon on lipolysis in liver [70]. It has been reported that a 14-day glucagon infusion decreases triglyceride content by 71% [85]. However, triglyceride accumulation was not observed in glucagon receptor knockout mice [77, 86]. Further investigation should be pursued to fully understand these observations.

C. Effects of Peptide Glucagon Receptor Antagonists

Hancock et al. had derived mice lacking pancreatic alpha cells to address the requirement for glucagon in the normal life of an animal. It was surprising that, despite the complete loss of glucagon-producing cells in the pancreas and no detectable circulating blood glucagon, affected animals are born alive and have a normal life span [87].

A number of peptide analogs of glucagon that incorporate amino acid substitutions and/or modifications have been synthesized to design pure glucagon antagonists by modifying the amino acid sequence of native glucagon [88]. These peptides block the action of glucagon by binding to the glucagon receptor without activating the associated signal transduction pathway.

Johnson DG et al. examined the ability of the glucagon analog [l-N alpha-trinitrophenylhistidine, 12-homoarginine]-glucagon (THG) to lower blood glucose concentrations in rats made diabetic with streptozotocin. They showed that THG is a potent antagonist of glucagon activation of the hepatic adenylate cyclase assay system.

Intravenous bolus injections of THG caused rapid decreases (20 to 35 percent) in blood glucose after 5 minutes. Continuous infusion of low concentrations of the inhibitor led to larger sustained decreases in blood glucose (30 to 65 percent). Their studies demonstrated that this glucagon receptor antagonist can substantially reduce blood glucose levels in diabetic animals without addition of exogenous insulin [89].

D. Effects of Non-Peptide Glucagon Receptor Antagonists

The continuous advances in peptide chemistry facilitated the discovery and characterization of an increasing number of more potent small-molecule non-peptide glucagon receptor antagonists [90-92]. Agents of this class usually bind with sites in the glucagon receptor that differ from the binding sites of glucagon. Skyrin, a bisanthroquinone that was identified as an active constituent of a fungal fermentation of a Penicillium sp., is an example of this group of glucagon receptor antagonists. Skyrin was found to act as a functional antagonist with respect to glucagon action [91]. Instead of inhibiting the binding of glucagon to its receptor, Skyrin seems to uncouple the receptor from adenylate cyclase activation [93]. Since the discovery of the inhibitory effects of Skyrin on glucagon receptor, pharmaceutical industry had identified more non-peptide glucagon antagonists with variable levels of glucose lowering effectiveness. Bay 27-9955 is another example of these non-peptide compounds that can be administered orally. Due its significant competitive glucagon antagonistic activity in humans, Bay 27-9955 was found to effectively block glucagon-induced increases in glucose production. [94].

MK-0893 was recently identified as a highly-selective glucagon receptor antagonist oral agent with a remarkable anti-hyperglycemic effect. The efficacy and safety of combination therapy of MK-0893 with Metformin or Sitagliptin was evaluated and presented during the 2011 European Association for the Study of Diabetes (EASD) meeting. T2DM patients either not on an antihyperglycemic agent with an HbA1c of 7.5-12%, or were on antihyperglycemic monotherapy or low-dose combination therapy with an HbA1c of 7-10% were eligible to be enrolled in the study. The primary endpoint was the change in 24-hr weighted mean glucose (WMG) from baseline after 4 weeks. Findings from this study displayed that substantial reduction in 24-hr WMG and fasting (FPG) and postprandial glucose was provided by all the treatment regimens at week 4. The MK-0893/metformin combination was superior to sitagliptin/metformin in lowering 24-hour weighted mean glucose, with a reduction of 117 mg/dL compared with 85 mg/dL. However, MK-0893/sitagliptin was significantly less effective than sitagliptin/metformin, which produced a 24-hour WMG reduction of 100 mg/dL. This could be because sitagliptin, a DPP-4 inhibitor, also suppresses glucagon and so there could be a threshold effect. Treatment with MK-0893 was well tolerated. However, there was a small but significant elevation in the liver enzymes alanine transaminase (ALT) and aspartate aminotransferase (AST) in the MK-0893/metformin group, and total cholesterol and LDL-cholesterol were increased from baseline with MK-0893/sitagliptin, relative to reductions with MK-0893/metformin and sitagliptin/metformin[95]

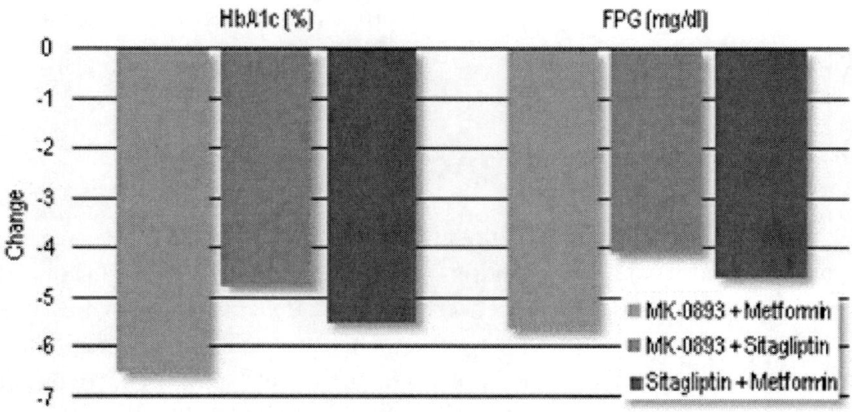

Figure 5. Change in hemoglobin A1c and fasting plasma glucose (FPG) levels after 4 weeks of treatment with combinations of MK-0893 40 mg o.d., sitagliptin 100 mg o.d. and metformin 1000 mg b.i.d. [Engel, S.S. et al., Abst 191: 47th European Association for the Study of Diabetes 2011 (EASD). September12-16,2011. Lisbon, Portugal. Figure is available online for public use: http://professional.diabetes.org/CongressReport_Display.aspx?CID=85652.

LY2409021 (LY) is another potent, selective antagonist of the glucagon receptor [96] Like MK-0893, LY2409021 produced statistically significant reductions in hemoglobin A_{1c} with small but significant elevations in liver enzymes [Information was presented in the European Association for the Study of Diabetes 47th Annual Meeting. September 13, 2011]. As indicated above, the potential risk of developing hypoglycemia and hyperglucagonemia needs to be carefully evaluated before these agents will be used for the treatment of diabetes. Due to the fact that alpha cell glucagon hypersecretion was shown to cause pancreatic tumors in animal studies, more research needs to be performed to evaluate this risk among subjects receiving glucagon receptor antagonists. Mu et al. evaluated the effects of Cpd-A, a potent and selective glucagon receptor antagonist, on alpha cells. Their work showed that chronic Cpd-A administration to mice resulted in sustained glucose lowering coincident with stable and moderate elevations in glucagon and GLP-1. Elevation of these hormones was reversible and did not result in a glucose overshoot upon termination of treatment. Furthermore, chronic treatment with Cpd-A for up to 82 days was not associated with overt changes in the morphology of pancreatic tissues [97]. Obviously, such duration of treatment is relatively short compared with the lifetime glucagon receptor blockage in Gcgr−/− mice. Although the moderate potency of Cpd-A might explain the lack of alpha cell hypertrophy, some significantly more potent molecules did not cause such change in alpha cells [unpublished data].

Conclusion

There is strong evidence that relative or absolute hyperglucagonemia plays a major role in the pathogenesis of type 2 diabetes. Both fasting and postprandial hyperglucagonemia were demonstrated in patients with type 2 diabetes. Furthermore, the glucagon response to arginine was found to be exaggerated. On the other hand, alpha cells of patients suffering from diabetes fail to respond appropriately to hypoglycemia. Targeting the glucagon action had become an important strategy in the treatment of diabetes. Currently, GLP-1 agonist therapy and DPP-4inhibitors are used to inhibit the excessive secretion of glucagon among patients suffering from diabetes. Other potential therapeutic modalities include the use of somatostatin analogs, glucagon antibodies, glucagon receptor gene inactivating agents and the administration of both peptide and non-peptide glucagon receptor antagonists.

References

[1] Rodriguez-Candela, J. L., P. M. Gofii, M. R. Caldeiro, and Y. P. Conzalez Carreras. 1947. Requerimiento de insulina antes y despues de la pancrectomit total, y previa ligadura de los conductos pancrekticos, en perros diabeticos por aloxana. *Trab. Inst. Nac. Cienc. Med* (Madrid). 9: 89.

[2] Assan R, Hautecouverture G, Guillemand S, Dauchy F, Protin P, De´ rot M. Evolution de parame` tres hormonaux (glucagon, cortisol, hormone somatotrope) et e´ nerge´ tiques (glucose, acides gras libres, glyce´ rol) dans dix acidoce´ toses diabe´ tiques graves traite´ es. *Pathol Biol* 1969; 17: 1095–1098.

[3] Unger RH, Aguilar-Parada E, Muller WA, Eisentraut AM. Studies of pancreatic alpha-cell function in normal and diabetic subjects. *J Clin Invest* 1970; 49: 837–848.

[4] Gerich JE, Lorenzi M, Bier DM, Schneider V, Tsalikian E, Karam JH, Forsham PH 1975 Prevention of human diabetic ketoacidosis by somatostatin. Evidence for an essential role of glucagon.*NEngl J Med* 292:985–989.

[5] Gerich JE, Schneider VS, Lorenzi M, Tsalikian E, Karam JH, Bier DM, Forsham PH 1976 Role of glucagon in human diabetic ketoacidosis: studies using somatostatin. *Clin Endocrinol* (Oxf) 5(Suppl):299S–305S.

[6] Stevenson RW, Williams PE, Cherrington AD. Role of glucagon suppression on gluconeogenesis during insulin treatment of the conscious diabetic dog. *Diabetologia* 1987;30:782–790

[7] Dinneen S, Alzaid A, Turk D, Rizza R. Failure of glucagon suppression contributes to postprandial hyperglycaemia in IDDM. *Diabetologia* 1995; 38:337–343

[8] Shah P, Vella A, Basu A, Basu R, Schwenk WF, Rizza RA. Lack of suppression of glucagon contributes to postprandial hyperglycemia in subjects with type 2 diabetes mellitus. *J Clin Endocrinol Metab* 2000;85:4053–4059

[9] Burcelin R, Katz EB, and Charron MJ. Molecular and cellular aspects of the glucagon receptor: role in diabetes and metabolism. *Diabetes Metab* 22: 373–396, 1996

[10] Baron AD, Schaeffer L, Shragg P, Kolterman OG. Role of hyperglucagonemia in maintenance of increased rates of hepatic glucose output in type II diabetics. *Diabetes* 1987; 36: 274–283.

[11] Reaven GM, Chen YD, Golay A, Swislocki AL, Jaspan JB. Documentation of hyperglucagonemia throughout the day in nonobese and obese patients with noninsulin-dependent diabetes mellitus. *J Clin Endocrinol Metab* 1987; 64: 106–110.

[12] Li, X. C., Liao, T. D. and Zhuo, J. L. (2008) Long-term hyperglucagonaemia induces early metabolic and renal phenotypes of Type 2 diabetes in mice. *Clin. Sci.* 114, 591–601

[13] Cryer PE. Glucagon and hyperglycaemia in diabetes. *Clin Sci (Lond).* 2008 May; 114(9):589-90.

[14] Ahren B, Larsson H. Impaired glucose tolerance (IGT) is associated with reduced insulin-induced suppression of glucagon concentrations. *Diabetologia* 2001; 44: 1998–2003.

[15] Larsson H, Ahren B. Islet dysfunction in insulin resistance involves impaired insulin secretion and increased glucagon secretion in postmenopausal women with impaired glucose tolerance. *Diabetes Care* 2000; 23:650–657.

[16] Mitrakou A, Kelley D, Mokan M et al. Role of reduced suppression of glucose production and diminished early insulin release in impaired glucose tolerance. *N Engl J Med* 1992; 326: 22–29.

[17] Mitrakou A, Kelley D, Veneman T, Jenssen T, Pangburn T, Rei, lly J, Gerich J 1990 Contribution of abnormal muscle and liver glucose metabolism to postprandial hyperglycemia in NIDDM. *Diabetes* 39:1381–1390.

[18] 18- Woerle HJ, Szoke E, Meyer C, Dostou JM, Wittlin SD, Gosmanov NR, Welle SL, Gerich JE 2006 Mechanisms for abnormal postprandial glucose metabolism in type 2 diabetes. *Am J Physiol Endocrinol Metab* 290:E67–E77.

[19] Ohneda A, Watanabe K, Horigome K et al. (1978) Abnormal response of pancreatic glucagon to glycemic changes in diabetes mellitus. *J Clin Endocrinol Metab* 46:504–510.

[20] Raskin P, Aydin I, Yamamoto T, Unger RH. Abnormal alpha cell function in human diabetes: the response to oral protein. *Am J Med* 1978; 64: 988–997.

[21] Raskin P, Aydin I, Unger RH. Effect of insulin on the exaggerated glucagon response to arginine stimulation in diabetes mellitus. *Diabetes* 1976; 25: 227–229.

[22] Aronoff SL, Bennett PH, Rushforth NB, Miller M, Unger RH: Arginine-stimulated hyperglucagonemia in diabetic Pima Indians. *Diabetes* 25:404–407, 1976.

[23] Aronoff SL, Bennett PH, Rushforth NB, Miller M, Unger RH: Normal glucagon response to arginine infusion in "prediabetic" Pima Indians. *J Clin Endocrinol Metab* 43:279 –286, 1976

[24] Raskin P, Aydin I, Unger RH: Effect of insulin on the exaggerated glucagon response to arginine stimulation in diabetes. *Diabetes* 25:227–229, 1976

[25] Gerich JE, Lorenzi M, Tsalikian E, Bohannon NV, Schneider V, Karam JH, Forsham PH: Effects of acute insulin withdrawal and administration on plasma glucagon responses to intravenous arginine in insulin-dependent subjects. *Diabetes* 25:955–960, 1976.

[26] Tsuchiyama N, Takamura T, Ando H, Sakurai M, Shimizu A, Kato K, Kurita S, Kaneko S. Possible role of alpha-cell insulin resistance in exaggerated glucagon responses to arginine in type 2 diabetes. *Diabetes Care* 30: 2583–2587, 2007.

[27] -Hamaguchi T, Fukushima H, Uehara M, Wada S, Shirotani T, Kishikawa H, Ichinose K, Yamaguchi K, Shichiri M: Abnormal

glucagon response to arginine and its normalization in obese hyperinsulinaemic patients with glucose intolerance: importance of insulin action on pancreatic alpha cells. *Diabetologia* 34:801–806, 1991.

[28] Rivera N, Ramnanan CJ, An Z et al. Insulin-induced hypoglycemia increases hepatic sensitivity to glucagon in a canine model. *J Clin Invest* 2010; 120: 4425–4435

[29] Jiang G, Zhang BB. Glucagon and regulation of glucose metabolism. *Am J Physiol Endocrinol Metab.* 2003;284: E671-E678.

[30] Gerich JE, Langlois M, Noacco C, Karam JH, Forsham PH. Lack of glucagon response to hypoglycemia in diabetes: evidence for an intrinsic pancreatic alpha cell defect. *Science* 1973; 182: 171–17.

[31] Cryer PE 2009 *Hypoglycemia in diabetes: pathophysiology, prevalence and prevention.* Alexandria, VA: American Diabetes Association; 45–95.

[32] Gerich J, Davis J, Lorenzi M, Rizza R, Bohannon N, Karam J, Lewis S, Kaplan R, Schultz T, Cryer P 1979 Hormonal mechanisms of recovery from insulin-induced hypoglycemia in man. *Am J Physiol* 236:E380–E38.

[33] Cryer, PE. Minireview: Glucagon in the Pathogenesis of Hypoglycemia and Hyperglycemia in Diabetes. *Endocrinology*, March 2012, 153(3):1039–1048.

[34] Segel SA, Paramore DS, Cryer PE 2002 Hypoglycemia-associated autonomic failure in advanced type 2 diabetes. *Diabetes* 51:724–733.

[35] White NH, Skor D, Cryer PE, Bier DM, Levandoski L, Santiago JV: Identification of type 1 diabetic patients at increased risk for hypoglycemia during intensive therapy. *N Engl J Med* 308:485–491, 1983.

[36] Foster DW, Rubenstein AH. Hypoglycemia. In*: Harrisons Principles of Internal Medicine.* Wilson JD, Braunwald E, Issel Bacher KJ, et al. (eds). McGraw-Hill Inc., New York. 1991, pp. 1759.

[37] Young CW, Karam JH. Hypoglycemic disorders. In: *Basic and Clinical Endocrinology. Greenspan* FS (ed). Appleton and Lange, East Norwalk. Connecticut 1991:651.

[38] Pampanelli S, Fanelli C, Lalli C, et al. Long term intensive insulin therapy: effects of HbA1c risk for severe and mild hypoglycemia, status of counter-regulation and unawareness of hypoglycemia. *Diabetologia* 1996;39: 677-86.

[39] Bolli GB, Perrielo G, Fanelli C, De Feo. Nocturnal blood glucose control in Type 1 diabetes mellitus. *Diabetes Care* 1993;3:71-89.

[40] Gerich JE, Langlois M, Noacco C, Karam JH, Forsham PH. Lack of glucagon response to hypoglycemia in diabetes. Evidence for an intrinsic pancreatic alpha cell defect. *Science* 1973; 182:171–3.

[41] Bolli G, De Feo P, Compagnucci P, Cartechini MG, Angeletti G, Santeusanio F, Brunetti P, Gerich JE. Abnormal glucose counterregulation in insulin-dependent diabetes mellitus: interaction of anti-insulin antibodies and impaired glucagon and epinephrine secretion. *Diabetes* 1983;32:134–41.

[42] Banarer S, McGregor VP, Cryer PE. Intraislet hyperinsulinemia prevents the glucagon response to hypoglycemia despite an intact autonomic response. *Diabetes* 2002;51:958–65.

[43] Raju B, Cryer PE. Loss of the decrement in intraislet insulin plausibly explains loss of the glucagon response to Hypoglycemia in Insulin-Deficient Diabetes. *Diabetes* 2005; 54:757–64.

[44] Bacatselos SO, Karamitsos DT, Kourtoglou GI, et al. Hypoglycemia unawareness in Type I diabetic patients under conventional insulin treatment. *Diab Nutr Metab* 1995;8:267–75

[45] Amiel SA, Sherwin RS, Simonson DC, Tamborlane WV. Effect of intensive insulin therapy on glycemic thresholds for counterregulatory hormone release. *Diabetes* 1988;37:901–7.

[46] Cryer PE: *Hypoglycemia: Pathophysiology, Diagnosis and Treatment.* Oxford University Press, New York, 1997, p. 85–125

[47] Fukuda M, Tanaka A, Tahar Y, Ikegami H, Yamamoto Y, Kumahara Y, Shima K: Correlation between minimal secretory capacity of pancreatic β-cells and stability of diabetic control. *Diabetes* 37:81–88, 1998.

[48] Dagogo-Jack SE, Craft S, Cryer PE: Hypoglycemia-associated autonomic failure in insulin dependent diabetes mellitus. *J Clin Invest* 91:819–828, 1993.

[49] Bagger J. I., Knop F. K., Holst J. J. & Vilsboll T. Glucagon antagonism as a potential therapeutic target in type 2 diabetes. *Diabetes, Obesity and Metabolism* 13: 965–971, 2011.

[50] Dobbs R, Sakurai H, Sasaki H, Faloona G, Valverde I, Baetens D,Orci L, Unger R 1975 Glucagon: role in the hyperglycemia of diabetes mellitus. *Science* 187:544–547

[51] Liljenquist JE, Bloomgarden ZT, Cherrington AD, Perry JM, Rabin D 1979 Possible mechanism by which somatostatin-induced glucagon suppression improves glucose tolerance during insulinopaenia in man. *Diabetologia* 17:139–143

[52] Gerich JE, Lorenzi M, Bier DM, Schneider V, Tsalikian E, Karam JH, Forsham PH 1975 Prevention of human diabetic ketoacidosis by somatostatin. Evidence for an essential role of glucagon. *N Engl J Med* 292:985–989

[53] Raskin P, Unger RH 1978 Hyperglucagonemia and its suppression. Importance in the metabolic control of diabetes. *N Engl J Med* 299:433–436

[54] Schusdziarra V, Rivier R, Dobbs RE, Brown M, Vale W,Unger RH (1978) Somatostatin analogs as glucagon suppressants in diabetes. *Horm Metab Res* 10:563-565

[55] Santeusanio F, Massi-Benedetti M, Angeletti G, Calabrese G, Bueti A, Brunetti P (1981) Glucagon and carbohydrate disorder in totally pancreatectomized man (A study with the aid of an artificial endocrine pancreas). *J Endocrinol Invest* 4:93-96.

[56] Flatt PR, Swanston-Flatt SK, Bailey CJ (1979) Glucagon antiserum: a tool to investigate the role of circulating glucagon in obese-hyperglycaemic (ob/ob) mice. *Biochem Soc Trans* 7:911-913 (Abstract)

[57] Unson CG, Gurzenda EM, Merrifield RB (1989) Biological activities of des-Hisl[Glu 9] glucagon amide, a glucagon antagonist. *Peptides* 10:1171-1177

[58] Thermos K, Meglasson MD, Nelson J, Lounsburry KM, Reisine T 1990. Pancreatic b-cell somatostatin receptors. *Am J Physiol* 259:E216–E224.

[59] Yamada Y, Kagimoto S, Kubota A, Yasuda K, Masuda K, Someya Y, Ihara Y, Li Q, Imura H, Seino S 1993 Cloning, functional expression and pharmacological characterization of a fourth (hSSTR4) and a fifth (hSSTR5) human somatostatin receptor subtype. *Biochem Biophys Res* Commun 195:844–852

[60] Hunyady B, Hipkin RW, Schonbrunn A, Mezey E 1997 Immunohistochemical localization of somatostatin receptor SST2A in the rat pancreas. *Endocrinology* 138:2632–2635

[61] Rossowski WJ, Coy DH 1994 Specific inhibition of rat pancreatic insulin or glucagon release by receptor-selective somatostatin analogs. *Biochem Biophys Res Commun* 205:341–346

[62] Reubi JC, Kappeler A, Waser B, Schonbrunn A, Laissue J 1998 Immunohistochemical localization of somatostatin receptor sst2A in human pancreatic islets. *J Clin Endocrinol Metab* 83:3746–3749.

[63] Vilsboll T, Krarup T, Madsbad S, Holst JJ. Defective amplification of the late phase insulin response to glucose by GIP in obese type II diabetic patients. *Diabetologia* 2002; 45: 1111–1119.

[64] Sorensen LB, Flint A, Raben A, Hartmann B, Holst JJ, Astrup A. No effect of physiological concentrations of glucagon-like peptide-2 on appetite and energy intake in normal weight subjects. *Int J Obes Relat Metab Disord* 2003; 27: 450–456.

[65] Meier JJ, Nauck MA, Pott A et al. Glucagon-like peptide 2 stimulates glucagon secretion, enhances lipid absorption, and inhibits gastric acid secretion in humans. *Gastroenterology* 2006; 130: 44–54.

[66] Meier JJ, Gallwitz B, Siepmann N et al. Gastric inhibitory polypeptide (GIP) dose-dependently stimulates glucagon secretion in healthy human subjects at euglycaemia. *Diabetologia* 2003; 46: 798–801.

[67] de Heer J, Rasmussen C, Coy DH, Holst JJ. Glucagon-like peptide-1, but not glucose-dependent insulinotropic peptide, inhibits glucagon secretion via somatostatin (receptor subtype 2) in the perfused rat pancreas. *Diabetologia* 2008; 51: 2263–2270.

[68] Brand CL, Rolin B, Jørgensen PN, Svendsen I, Kristensen JS, Holst JJ 1994 Immunoneutralization of endogenous glucagon with monoclonal glucagon antibody normalizes hyperglycaemia in moderately streptozotocin-diabetic rats. *Diabetologia* 37:985–993

[69] Jelinek LJ, Lok S, Rosenberg GB, Smith RA, Grant FJ, Biggs S, Bensch PA, Kuijper JL, Sheppard PO, Sprecher CA, O'Hara PJ, Foster D, Walker KM, Chen LHJ, McKernan PA, Kindsvogel W: Expression cloning and signaling properties of the rat glucagon receptor. *Science* 259:1614–1616, 1993.

[70] Lok S, Kuijper LJ, Kramer JM, Whitmore TE, Specher CA, Mathews S, Grant FJ, Biggs SH, Rosenberg GB, Sheppard PO, O'Hara PJ, Foster DC, Kingsvogel W: The human glucagon receptor encoding gene: structure, cDNA sequence and chromosomal localization. *Gene* 140:203–209, 1994.

[71] Jiang G, Zhang BB. Glucagon and regulation of glucose metabolism. *AmJ Physiol Endocrinol Metab* 2003; 284: E671–E678.

[72] Boden G, Tappy L, Jadali F, Hoeldtke RD, Rezvani I, Owen OE: Role of glucagon in disposal of an amino acid load. *Am J Physiol* 259:E225_E232,1992.

[73] Pilkis SJ, Granner DK: Molecular physiology of the regulation of hepatic gluconeogenesis and glycolysis. *Annu Rev Physiol* 54:885–909, 1992.

[74] Bagger J. I., Knop F. K., Holst J. J. & Vilsboll T. Glucagon antagonism as a potential therapeutic target in type 2 diabetes. *Diabetes, Obesity and Metabolism* 13: 965–971, 2011

[75] Hansen LH, Abrahamsen N, Nishimura E. Glucagon receptor mRNA distribution in rat tissues. *Peptides* 1995; 16: 1163–1166.

[76] Ali S, Drucker DJ. Benefits and limitations of reducing glucagon action for the treatment of type 2 diabetes. *Am J Physiol Endocrinol Metab* 2009; 296: E415–E421

[77] Parker JC, Andrews KM, Allen MR, Stock JL, McNeish JD 2002 Glycemic control in mice with targeted disruption of the glucagon receptor gene. *Biochem Biophys Res Commun* 290:839–843

[78] Gelling RW, Du XQ, Dichmann DS, Romer J, Huang H, Cui L, Obici S, Tang B, Holst JJ, Fledelius C, Johansen PB, Rossetti L, Jelicks LA, Serup P, Nishimura E, Charron MJ 2003 Lower blood glucose, hyperglucagonemia, and pancreatic _-cell hyperplasia in glucagon receptor knockout mice. *Proc Natl Acad Sci USA* 100:1438–1443

[79] Zhou C, Dhall D, Nissen N, Chen C, Yu R. Homozygous P86S mutation of the human glucagon receptor is associated with hyperglucagonemia, alpha cell hyperplasia, and islet cell tumor. *Pancreas* 2009; 38: 941–946

[80] Hansen L, Larger E, Chaput J et al. Identification of a glucagon receptor gene deletion mutation in a patient with hyperglucagonemia and pseudoadenomatous hyperplasia of pancreatic alpha-cells. *J Clin EndocrinolMetab* 1999; 84: P3–P441

[81] Sloop KW, Cao JX, Siesky AM, Zhang HY, Bodenmiller DM, Cox AL, Jacobs SJ, Moyers JS, Owens RA, Showalter AD, Brenner MB, Raap A, Gromada J, Berridge BR, Monteith DK, Porksen N, McKay RA, Monia BP, Bhanot S, Watts LM, Michael MD 2004. Hepatic and glucagon-like peptide-1-mediated reversal of diabetes by glucagon receptor antisense oligonucleotide inhibitors. *J Clin Invest* 113:1571–1581.

[82] Liang Y, Osborne MC, Monia BP, Bhanot S, Gaarde WA, Reed C, She P, Jetton TL, Demarest KT 2004 Reduction in glucagon receptor expression by an antisense oligonucleotide ameliorates diabetic syndrome in db/db mice. *Diabetes* 53:410–417

[83] Butler M, McKay RA, Popoff IJ, Gaarde WA, Witchell D, Murray SF, Dean NM, Bhanot S, Monia BP: Specific inhibition of PTEN expression reverses hyperglycemia in diabetic mice. *Diabetes* 51:1028–1034, 2002.

[84] Gum R, Gaede L, Koterski S, Heindel M, Clampit J, Zinker B, Trevillyan J, Ulrich R, Jirousek M, Rondinone C: Reduction of protein tyrosine phosphatase 1B increases insulin-dependent signaling in ob/ob mice. *Diabetes* 52:21–28, 2003.

[85] She P, Shiota M, Shelton K, Chalkley R, Postic C, Magnuson M: Phosphoenolpyruvate carboxykinase is necessary for the integration of hepatic energy metabolism. *Mol Cell Biol* 20:6508–6517, 2000

[86] Hippen A, She P, Young J, Beitz D, Lindberg G, Richardson L, Tucker R: Alleviation of fatty liver in dairy cows with 14-day intravenous infusions of glucagon. *J Dairy Sci* 82:1139–1152, 1999

[87] Amatruda J, Livingston J: The search for glucagon antabonists. In *Handbook of Experimental Pharmacology*. Lefebvre PJ, Ed. Berlin, Springer, 1996, p. 133–147

[88] Hancock AS, Du A, Liu j, Miller M, May C. Glucagon Deficiency Reduces Hepatic Glucose Production and Improves Glucose Tolerance In *Adult Mice. Mol Endocrinol*, August 2010, 24(8):1605–1614.

[89] Azizeh BY, Van Tine BA, Trivedi D, Hruby VJ: Pure glucagon antagonists: biological activities and cAMP accumulation using phosphodiesterase inhibitors. *Peptides* 18:633–641, 1997.

[90] Johnson DG, Goebel CU, Hruby VJ, Bregman MD, Trivedi D. Hyperglycemia of diabetic rats decreased by a glucagon receptor antagonist. *Science* 1982; 215: 1115–1116]

[91] Van Tine BA, Azizeh BY, Trivedi D et al. Low level cyclic adenosine 3',5'-monophosphate accumulation analysis of [des-His1, des- Phe6 Glu9] glucagon-NH2 identifies glucagon antagonists from weak partial agonists/antagonists. *Endocrinology* 1996; 137: 3316–3322.

[92] Parker JC, McPherson RK, Andrews KM et al. Effects of skyrin, a receptorselective glucagon antagonist, in rat and human hepatocytes. *Diabetes* 2000; 49: 2079–2086.

[93] Gysin B, Johnson DG, Trivedi D, Hruby VJ. Synthesis of two glucagon antagonists: receptor binding, adenylate cyclase, and effects on blood plasma glucose levels. *J Med Chem* 1987; 30: 1409–1415.

[94] West RR, Smith RA, McKernan PA, Chen LHJ, Rosenberg GB, Biggs S, Delmendo R, Pedigo L, Piggott JR: Novel Pharmacological and Agrobiological Activities. *Third International Conference on the Biotechnology of Microbial Products. Rohnert Park, CA,* 1993, p. 45

[95] Petersen KF, Sullivan JT. Effects of a novel glucagon receptor antagonist (Bay 27-9955) on glucagon-stimulated glucose production in humans. *Diabetologia* 2001; 44: 2018–2024.

[96] Engel, S.S. et al., Abstract 191. *47th European Association for the Study of Diabetes 2011* (EASD), September 12 - 16, 2011. Lisbon, Portugal].

[97] Ronan P. Kelly, Eyas J. Abu-Raddad, Lai San Tham, Haoda Fu, Jane A. Pinaire, Mark A. Deeg. *Single Doses of the Glucagon Receptor*

Antagonist LY2409021 Reduce Blood Glucose in Healthy Subjects and Patients with Type 2 Diabetes Mellitus. Abstract Number: 1004-P. ADA 71st Scientific Sessions (2011) June 24 - 28, 2011. San Diego Convention Center - San Diego, California.

[98] Mu J, Jiang G, Brady E, Dallas-Yang Q, Liu F, Woods J, Zycband E, Wright M, Li Z, Lu K, Zhu L, Shen X, Sinharoy R, Candelore ML, Qureshi SA, Shen DM, Zhang F, Parmee ER, Zhang BB. Chronic treatment with a glucagon receptor antagonist lowers glucose and moderately raises circulating glucagon and glucagon-like peptide 1 without severe alpha cell hypertrophy in diet-induced obese mice. *Diabetologia*. 2011 Sep;54(9):2381-91

In: Glucagon
Editor: Ahmed Al-Jebawi

ISBN: 978-1-62618-147-2

Chapter 5

The Role of Incretin Hormones in Glucagon Regulation and Diabetes Treatment

Shushan B. Artinian, Sawsan M. Al-Lafi, Suzan S. Boutary, Nadine S. Zwainy and Anwar B. Bikhazi *

Department of Anatomy, Cell Biology and Physiological Sciences, Faculty of Medicine, American University of Beirut, Beirut, Lebanon

Abstract

Incretins are hormones that are secreted by the gut upon feeding; their release alerts the pancreatic islets that nutrients will come through the gastrointestinal tract and islets start priming by vagal stimulation (Boron and Boulpaep, 2005). Glucagon-Like Peptide-1 (GLP-1) and Glucose Dependent Insulinotropic Peptide (GIP) are the major incretin hormones produced by the L cells of the ileum and colon, and K cells of the duodenum and jejunum, respectively (Xu et al., 2007). GLP-1 has numerous postprandial effects on glucose homeostasis. Upon interaction with its receptor on β-cells, GLP-1 increases the intracellular levels of

* Corresponding Author: Anwar B. Bikhazi, PhD, ab04@aub.edu.lb, Department of Anatomy, Cell Biology and Physiological Sciences, Faculty of Medicine, American University of Beirut, PO Box 11-0236, Riad El-Solh 1107 2020. Beirut, Lebanon.

cAMP and calcium, thereby releasing insulin. GLP-1 has also inhibitory effects on Glucagon secretion from the α-cells, gastric emptying and food ingestion (Drucker, 2007). In both type 1 and type 2 diabetes mellitus, there is a marked reduction in the incretin effect (Knop et al., 2007). A recent study showed that GLP-1 receptor expression is down regulated in β-cells exposed to high glucose concentrations in vitro and hyperglycemia in vivo (Xu et al., 2007). The β-cells in diabetic individuals exhibit attenuated sensitivity to GLP-1. Moreover, Type 2 diabetic patients exhibited significantly lower GLP-1 secretion in response to meal absorption (D'Alessio and Vahl, 2004). This chapter, focuses on the function of incretin hormones, specifically, GLP-1 and its analogues on glucagon homeostasis and diabetes.

Glucose Homeostasis

Glucose homeostasis plays an important role in the normal physiology of the body. In a healthy person, fasting blood glucose concentration remains below 5.5 mmol/L (100 mg/dl). On the other hand, postprandial glucose levels should not exceed 140 mg/dl (7.8 mmol/L). The pancreas is the key organ that is responsible for keeping blood glucose levels within the normal range. The pancreas is both an exocrine and an endocrine gland that produces digestive enzymes and hormones responsible for regulation of fuel use, respectively (Rhoades and Bell, 2008). In 1869, Paul Langerhans discovered a cluster of cells in the endocrine pancreas that had unique morphological features (Eknoyan and Nagy 2005). These Islets of Langerhans, which were later named after him, are composed of 6 different cell types: α, β, δ, PP, D1 and enterochromaffin cells. The two major cell types are the β-cells that synthesize and secrete insulin and α-cells that produce and release glucagon (Seifter et al., 2005). These cells coordinate and work antagonistically in response to blood glucose levels. When glucose level increases, the β-cells secrete insulin and the α-cells reduce their release of glucagon. On the other hand, when hypoglycemia occurs, β-cells halt their secretion of insulin and α-cells start releasing glucagon (Quesada et al., 2008).

Insulin

Insulin was the first pancreatic hormone identified by Banting and Best in 1922. It was first extracted from cow (bovine) and later from pig (porcine).

Bovine and porcine insulin worked very well (and still do) for the majority of patients, but some could develop an allergy or other types of reactions to the foreign protein. In the 1980's, technology had advanced to the point where human insulin could be made. The advantage would be that human insulin would have a much lower chance of inducing a reaction because it is not a foreign protein. This was made possible by the development of recombinant DNA techniques in which the human gene that codes for the insulin protein was cloned and then put inside of bacteria (Bowen, Austgen, and Rouge, 2006). Insulin is a hormone secreted from the pancreas composed of two peptide chains A and B that are linked together by two disulfide bonds with an additional disulfide within the A chain. In most species, the A chain consists of 21 amino acids and the B chain of 30 amino acids. Although the amino acid sequence of insulin varies among species, certain segments of the molecule are highly conserved, including the positions of the three disulfide bonds, both ends of the A chain and the C-terminal residues of the B chain (Bowen, Austgen, and Rouge, 2006). These similarities in the amino acid sequence of insulin lead to a three dimensional conformation of insulin that is very similar among species. Insulin from one animal is often biologically active in other species; indeed pig insulin has been widely used to treat human patients.

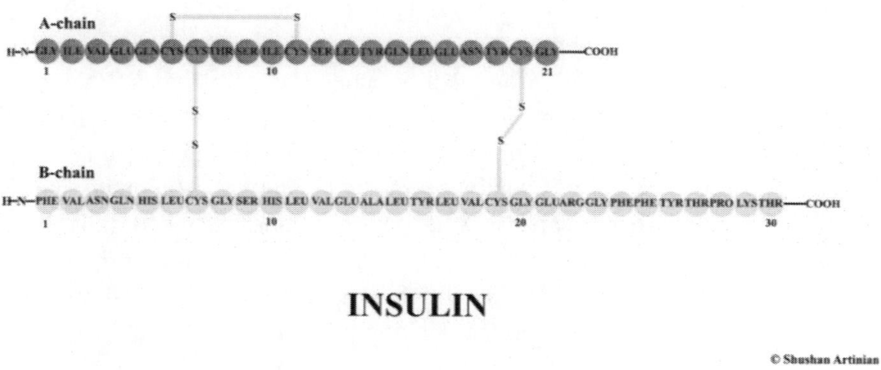

Figure 1. Molecular Structure of the 51 amino acid Insulin.

Insulin is only produced in the pancreatic β cells. The insulin gene found on chromosome 11's short arm is responsible for producing the mRNA that encodes preproinsulin. This preprohormone is later cleaved in the Endoplasmic Reticulum forming the proinsulin that consists of chains A, B and C. It is in the golgi apparatus, trans-golgi to be precise, where proinsulin is cut at two specific sites that separates chains A from B and B from C. Later, chains A and B are joined by 2 disulfide bridges rendering the insulin an active 51 amino acid molecule (Figure 1). Insulin and free C peptide molecules are packaged in the Golgi apparatus into secretory granules that accumulate in the cytoplasm. When pancreatic β-cell is stimulated by high glucose levels, insulin is secreted from the cell by exocytosis and diffuses into islet capillary blood. The C peptide is also secreted into the blood stream, but has no known biological activity. (Boron and Boulpaep, 2005)

Actions of insulin are abundant. Insulin acts to regulate glucose homeostasis (Pessin and Saltiel, 2000). As blood glucose levels increase, insulin is secreted to ensure the disposal of excess glucose molecules from the circulation. This is done by recruiting glucose transporters (GLUT4) to the muscular and adiposal membranes that facilitate the internalization of glucose molecules into the cell (Rhoades and Bell, 2008). Moreover, insulin clears the blood of excess glucose by promoting glycogen synthesis in the liver through the activation of glycogen synthase (Boron and Boulpaep. 2005). In lipid metabolism, it promotes the storing of fats, by blocking the action of fat releasing lipase in adipocytes (Seifter et al., 2005). Furthermore, this potent anabolic hormone plays an important role in growth and the development of tissues. (Pessin and Saltiel, 2000). The main target organs of insulin are: the liver, the skeletal muscle and the adipose tissue (Rhoades and Bell, 2008). The insulin receptor is a member of the tyrosine kinase receptor family. It is a dimeric protein made up of 2 α and 2 β domains (De Meyts and Whittaker, 2002). Insulin receptors are typically expressed in hepatocytes, adipocytes and muscle cells, since those are the insulin responsive tissues (Beauchamp et al, 2010). Upon the binding of insulin to the extracellular α domain, many intracellular changes occur: the tyrosine kinase domain of the β subunit is activated (Pessin and Saltiel, 2000). This follows the phosphorylation of insulin-receptor substrates (IRS-1 and IRS-2) and Src homology domain 2 (SHC), the latter is able to bind to the docking proteins (IRS-1 and 2). Hence, a signaling transduction cascade is activated that involves GRB2 and Ras protein that eventually triggers MAP and MEK kinases and enhances proliferation and gene expression (Boron and Boulpaep, 2005).

While insulin acts anabolically inducing cells to take up glucose circulating in the blood and store them as glycogen and fat, glucagon on the other hand acts catabolically, by initiating liver gluconeogenesis and glycogenolysis which consequently releases glucose into the bloodstream (Quesada et al., 2008).

Glucagon

Although hypoglycemia is the main stimulant for the secretion of glucagon, the autonomic nervous system and amino acids also have the capability to induce alpha cell secretion. Alpha cells express the proglucagon gene and hence produce Glucagon. Glucagon is the primary hormone regulating hepatic glucose production and ketogenesis. It is a 29 amino acid peptide that is produced from the cleavage of preproglucagon (Figure 2).

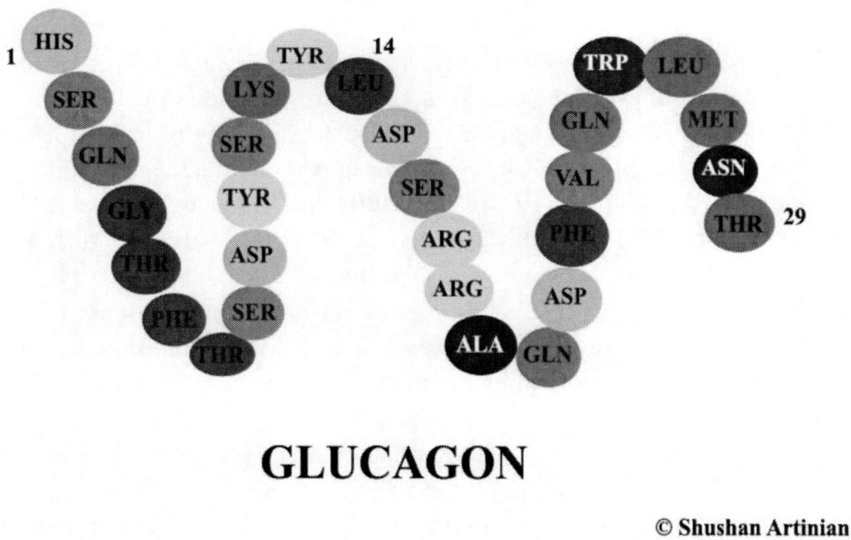

Figure 2. Molecular Structure of the 29 amino acid Glucagon.

Glucagon receptor is a glycoprotein that is present on hepatic plasma membrane. It activates glycogen phosphorylase enzyme, which is responsible for the breakdown of glycogen into glucose-1-phosphate. As glucagon is released, it binds to a G protein coupled receptor, which activates G-protein mediated signal transduction. Consequently, Gsα activates adenylate cyclase enzyme, by which cAMP production is induced. The rise in cAMP increases protein kinase activity. Gqα is also activated and results in an increase of Ca^{++} ions by phospholipase C. Upon the glucagon receptor activation, both gluconeogenesis and glycogenolysis processes are promoted, while glycogenesis is inhibited. (Juul Hare, 2010)

Insulin and glucagon interplay to maintain a constant glucose level in blood. The molar ratio of insulin to glucagon is approximately 2.0. Following starvation, insulin to glucagon ratio drops down to 0.5, which results in the mobilization of amino acids and promotes increased glycogenolysis. Conversely, a glucose rich meal induces a rise in this ratio, which may reach 10 or more and as a consequence enhance glucose uptake, oxidation and its conversion to glycogen in the liver and muscles (Berne et al., 2004).

Diabetes Mellitus

Diabetes Mellitus is a heterogeneous metabolic disorder characterized by high blood glucose levels due to defects in insulin secretion or action (Alberti et al., 1998; Holt 2004; Noyan et al., 2005; Robles and Singh-franco, 2009). If untreated, Diabetes Mellitus causes improper carbohydrate, fat and protein metabolism (Alberti et al., 1998; Holt 2004) that may lead to polydipsia, polyuria, weight loss, blurred vision, coma or death (Seifter et al., 2005, Rhoades and Bell, 2009). Moreover, Diabetes Mellitus shows two types of complications namely, acute and chronic. Acute complications include hyperglycemia, ketoacidosis and fluid loss. The chronic complications affect the circulatory system thus leading to micro and macrovascular complications. The microvascular complications are characterized by disturbances of blood flow in small blood vessels leading to retinopathy, nephropathy and neuropathy. The macrovascular complications on the other hand, lead to coronary vascular disease, peripheral vascular disease and stroke (Rhoades and Bell, 2009).

In 1980 the World Health Organization (WHO) classified DM according to the need of insulin into: (1) insulin dependent diabetes mellitus (IDDM) or type 1 diabetes and (2) non-insulin dependent diabetes mellitus (NIDDM) or type 2 diabetes mellitus. In 1985 type-1 and type-2 terms were omitted. In the year 1997 however, the American Diabetes Association (ADA) declared a new classification according to the etiology of the disease: type-1 and type-2 diabetes mellitus. These two terms are still in use today (Alberti et al., 1998; Holt, 2004).

Type-1 Diabetes Mellitus is an autoimmune disorder that results from the destruction of the pancreatic beta cells by the immune system (Seifter et al., 2005; Rhoades and Bell, 2009). It is characterized by high blood glucose levels, known as hyperglycemia, due to insufficient insulin production. It represents around 10% of all diabetic cases worldwide (Holt, 2004, Seifter et al., 2005). Although type 1 diabetes affects all age groups, the majority of individuals are diagnosed either at around the age of 4–5 years, or in their teens and early adulthood (Holt, 2004).

Recently, type-1 Diabetes Mellitus was further classified into type-1A and type-1B. Type-1A is immune-mediated while type-1B is the result of inflammation, since histological studies done on pancreatic tissue showed no auto antibodies (Rhoades and Bell, 2009). Moreover, the main cause of type-1 Diabetes Mellitus is not fully known or understood, but it is speculated that genetic predisposition and several environmental factors, such as viral infection, nutritional factors and low birth weight, play an important role in its prevalence (Holt, 2004). Patients with type-1 DM should be treated with insulin therapy to control the development of hyperglycemia, glucosuria, dehydration, polyuria and ketoacidosis. Failure to replace insulin in type 1 diabetic patients usually leads to a decrease in the fluid volume, which in turn reduces the circulatory blood volume and increases the production of ketones that lead to acidosis, loss of cations, coma and death (Rhoades and Bell, 2009).

Type-2 Diabetes Mellitus is an epidemic disease that accounts for around 90% of all diabetic cases worldwide (Rhoades and Bell, 2009). It usually affects adults. The WHO speculated that the number of adults affected will double from 177 million in 2003 to 370 million in 2030. As type-2 diabetes Mellitus develops, insulin resistance progresses causing hyperinsulinemia and further proceeds to insulin deficiency. Genetic predisposition and environmental factors such as obesity and physical inactivity are important factors for the development of this disease (Holt 2004). Patients with type-2 Diabetes Mellitus exhibit hyperglycemia that may result in fluid loss and dehydration. They are usually obese and suffer from insulin resistance and their treatment includes weight loss, diet and exercise. Many patients with type 2 diabetes receive treatment with oral anti-hyperglycemic agents along lifestyle modifications. Based on their mechanism of action, oral agents are classified into insulin sensitizers, insulin secretagogues, alpha glucosidase inhibitors and dipeptidyl peptidase-4 inhibitors (Rhoades and Bell, 2009).

Treatment Modalities for Diabetes Mellitus

Insulin has long been used as treatment for diabetic patients, however it is associated with weight gain (Hinnen, Nielsen, Waninger, and Kushner, 2006).

Now, new treatment modalities are under intensive study. The etiology of diabetes as well as its complications, such as cardiovascular problems, are taken into consideration when treating diabetic patients. The more recent modalities in treatment include the use of pharmacological agents similar in their action to incretins secreted by gut mucosa. The incretin hormone is defined by two characteristics: first it is released after nutrient ingestion, and second the concentrations of glucose reached after meal ingestion are enough to stimulate its secretion (Porte et al., 2003). Our body has the capacity to adjust its glycemic levels. After nutrients make their way to the stomach and through the gastrointestinal (GI) tract, there will be triggering of hormonal and neural signals that will control nutrient absorption, gut motility, gastric emptying, and energy storage (Drucker, 2003). The GI mucosal lining is the early integrator of information about nutrient glucose concentration and storage. It plays a major role by secreting insulinotropic peptides that connect insulin secretion to substrate absorption. These were called incretin hormones because they stimulate pancreatic endocrine secretions in contrast to excretins that stimulate pancreatic exocrine secretions. The importance of incretin hormones was recognized when the circulating level of insulin was 30-60% higher among normal subjects who received the glucose orally than those who received glucose intravenously; this is termed incretin effects (Poretsky, 2002). However, in diabetic patients there is minimal augmentation of insulin release with oral versus intravenous glucose supplementation because of the defects in diabetic incretin hormone (Porte et al., 2003).

Incretin Hormones

Two hormones fall under the criteria of incretin hormones and are under intensive research including glucose-dependent insulinotropic polypeptide (GIP) and glucagon-like peptide-1 (GLP-1). Recent studies suggest that both hormones are important for glucose control (Berghofer, Peterson, Schneider, and Fehmann, 1997).

When glucose is intravenously infused at a constant rate, a biphasic insulin secretory response results, displaying a rapid rise of a peak followed by a slower second peak. In contrast, the administration of glucose orally, triggers a hormonal pathway which eventually leads to a far greater response of insulin secretion that can last as long as glucose is administered. This phenomenon is called the 'Incretin Effect' (Rhoades and Bell, 2008).

Upon ingestion, incretin hormone secretion by the gut alerts the pancreatic islets of the arrival of nutrients through the gastrointestinal tract; hence islets start priming by vagal stimulation (Boron and Boulpaep, 2005). Glucagon-Like Peptide-1 (GLP-1) and Glucose Dependent Insulinotropic Peptide (GIP) are the major incretin hormones produced by the L cells of the ileum and colon, and K cells of the duodenum and jejunum, respectively (Inzucchi and McGuire, 2008).

The proglucagon gene, located on chromosome 2 in humans, has the coding sequence of GLP-1. Pancreatic α-cells, intestinal L-cells and neural cells in the caudal brainstem and hypothalamus, express the proglucagon gene (Baggio and Drucker, 2007) and the mRNAs produced in the pancreas and intestine are identical. Its post-translational processing, however, differs in the two tissues. The post translational processing of proglucagon in pancreatic α-cells results in Glucagon, Glicentin-Related Pancreatic Polypeptide (GRPP), Intervening peptide-1 (IP-1) and the major proglucagon fragment (MPGF) (Holst, 2007). The latter codes for the production of GLP-1 (1-36 amide), GLP-1 (1-37) and GLP-2 in the pancreas (Hui et al., 2005). Whereas the intestinal L and brain cells produce GLP-1 (7-36 amide), GLP-1 (7-37), GLP-2, IP-2, and glicentin that if further cleaved produces GRPP and oxyntomodulin (Holst, 2007). The enzyme prohormone convertase (PC) 1/3 cleaves proglucagon to generate GLP-1 (Shin et al., 2008).

Glucose –Dependent Insulinotropic Polypeptide

Glucose-Dependent Insulinotropic Polypeptide (GIP) is a 42-amino acid peptide secreted by K cells that are present in the mucosa of duodenum and jejunum (Drucker, 2003). Its secretion is highly dependent on glucose concentration (>5-6 mmol). The GIP amino acid sequence is highly conserved among species. In normal humans, it is found to stimulate insulin secretion in the same way as what it does post-prandially; while in diabetic rats it was found that GIP levels are increased suggesting a role of incretin hormones in the pathophysiology of diabetes. In type 2 diabetes, it is thought that GIP receptor undergoes desensitization. However, if glucose was given intravenously, no increase in the level of GIP was observed suggesting that the interaction between substrate and gut mucosa is essential to induce GIP secretion. One GIP receptor has been identified on α and β pancreatic islet, upper GI tract, adrenal cortex, pituitary, and adipocytes. When ligand binds to the receptor, it triggers cyclic AMP secretion that will induce depolarization-stimulated exocytosis from β-cells. GIP is thought to reduce gastric emptying via the stimulation of gastric somatostatin release (Porte, Sherwin, and Baron, 2003).

Glucagon-Like Peptide-1

Glucagon-Like Peptide-1 (GLP-1) is an incretin hormone produced primarily by the L-cells in the mucosa of the ileum and the colon (Mannucci and Rotella, 2008).

Small neurons found in the Nucleus Tractus Solitarius (NTS), caudal brainstem, also produce GLP-1 that plays the role of a neuromodulator (Berthoud, 2009). The central nucleus of the amygdala (CeA) and the paraventricular nucleus of the hypothalamus (PVN) are also sites of GLP-1 production (Kinzig et al., 2002).

Proglucagon is the precursor of GLP-1 (Drucker, 2003). It is secreted by pancreatic α-cells, the L cells located in the intestinal mucosa, and a population of neurons located in the hind brain as a result of ingestion of fats and carbohydrates.

Proglucagon is then cleaved in the pancreas by prohormone convertase 1/3 (PC1/3), only found in α-cells, and prohormone convertase 2 (PC2) to yield GLP-1 and GLP-2 after nutrient ingestion (Bowen, Austgen, and Rouge, 2000; Wei and Mojsov, 1995). Two forms of GLP-1 have been identified: GLP-1(7-37) and GLP-1(7-36) that differ by only single amino acid. However, around 80% of circulating GLP-1 is in the form GLP-1 (7-36) (Drucker, 2003) (Figure 3).

GLP-1 producing L-cells are found predominantly in the ileum, colon and more proximally in duodenum and jejunum (Drucker, 2003). GLP-1 release is rapidly stimulated by oral ingestion of a meal, but not with intravenous glucose infusion. Its secretion is independent of the amount of the meal ingested, but rather on the rate of nutrient absorption. GLP-1 is thought to respond to oral food ingestion by neural message. Dipeptidyl peptidase-IV cleaves the N-terminal His-Ala from GLP-1 (7-36) forming GLP-1 (9-36). Since N-terminal of GLP-1 is important for its action; cleaving N-terminal by DPP-IV inactivates GLP-1 (Lovshin and Drucker, 2000).

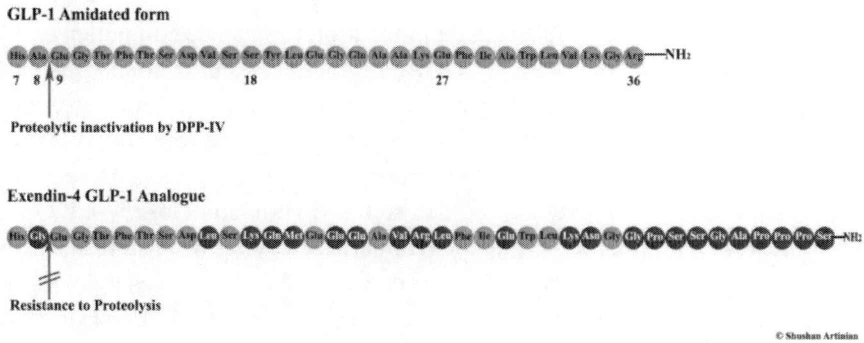

Figure 3. Molecular Structures of Active GLP-1 (7-36) and its analogue Exendin-4.

GLP-1 receptor is expressed in the islet cells, kidney, lung, brain, heart, and gastrointestinal tract (Gros et al., 2006). GLP-1 receptor is a G-protein-coupled receptor superfamily that can couple to Gαs-containing heterotrimeric G-proteins. It is a member of vasointestinal peptides (VIP) (Gomez, Pritchard, and Herbert, 2002; Bullock, Heller, and Habener, 1996). GLP-1 acts by binding to its receptor to mediate β-cell insulin secretion through activation of adenyl cyclase, formation of intracellular cyclic AMP (Mattson, Perry, and Greig, 2003), and activation of protein kinase A (PKA) (D'Alessio and Vahl, 2004). This will result in closure of voltage-gated potassium channels (that are activated in case of hyperglycemia), increase in intracellular calcium, and granular exocytosis at site distal to elevation of intracellular calcium. Also a number of kinases are activated upon the binding of GLP-1 to its receptor including Phosphatidylinositol 3-kinase (PI3-kinase) and extracellular regulated kinase (Erk) (Gomez, Pritchard, and Herbert, 2002). Activation of GLP-1 receptor improves β-cell function and survival following induction of endoplasmic reticulum stress in diabetes (Yusta et al., 2006).

GLP-1 Therapy

The effect of GLP-1 on insulin secretion is dependent on glucose level. GLP-1 stimulates insulin secretion when glucose levels exceed 5 mmol/l. Administration of GLP-1 induces insulin secretion and biosynthesis, inhibits β-cell apoptosis, and stimulates transcription of β-cell genes including glucokinase and GLUT-2 glucose transporter through the phosphatidylinositol

3-kinase pathway (Mojsov, 2000; Suzuki et al., 2003). It also activates glucose-insensitive β-cells; this is called induction of glucose competence.

Recent studies in rats with β-cell damage after streptozotocin treatment suggest a role of α-cell-derived GLP-1 in the promotion of insulin secretion (D'Alessio and Vahl, 2004). Like β-cells, α-cells seem to be tonically regulated by GLP-1. GLP-1 stimulates insulin secretion and inhibits glucagon secretion with the help of insulin and/or somatostatin (Fehmann, Goke, and Goke, 1992). GLP-1 also affects the gastrointestinal tract by slowing its transit and decreasing secretion of enzymes (Gautier et al., 2005). It increases pyloric tone and inhibits antral contractions, thus lowering rate of gastric emptying and slowing nutrient absorption of both liquid and solid meals in fed, but not in the fasting state ; these were called 'ileal brake' effect (Mojsov, 2000; Holst, 2005). Its action is thought to be due to vagal afferent signal from the gastrointestinal lining. So it plays a regulatory role in satiety by working on cerebroventricular system and body weight (Porte et al., 2003).

GLP-1 is one of the most promising antidiabetic drugs and it is under intensive research studies. It stimulates insulin secretion and neogenesis of pancreatic islets, and slows apoptosis of pancreatic cells (Takei and Kasatani, 2004; Holst, 2005). Some say that GLP-1 secretion increases in diabetes and others report the opposite, but they all agree on the importance of GLP-1 in regulating the glucose levels among patients who suffer from diabetes. (Lugari et al., 2002; Porte et al., 2003).

Glucagon Receptor Antagonist

On the other hand, suppression of glucagon has also been considered as a treatment for diabetes mellitus. In fact, several agents including Skyrin, and NNC 25-0926 have been heavily studied. Skyrin, a receptor-selective glucagon antagonist, is a fungal bisanthroquinone, which acts mainly by inhibiting glucagon stimulated cAMP production and glycogenolysis. This in turn markedly reduces glucose output in the plasma. Moreover, this effect was specific for glucagon receptor, since researchers did not observe any effect on Glucagon like Peptide-1 stimulated cAMP production. The mechanism by which Skyrin acts on the Glucagon receptor however, remains unclear. It is assumed that Skyrin may either act through disrupting the communication of the glucagon receptor with its G-Protein proteins or it may be that skyrin works at a secondary site, possibly interfering at the transmembrane to reduce the affinity of glucagon to its receptor. (Parker et al., 2000).

NNC 25-0926 which is a (2*R*)-*N*-[4-({4-(1-cyclohexen-1-yl) [(3,5-dichloroanilino) carbonyl] anilino} methyl)benzoyl]-2-hydroxy-*b*-alanine has been assessed as a glucagon receptor antagonist. It is a potent, competitive molecule which readily binds to the cloned human glucagon receptor expressed in baby hamster kidney cells with a half maximal inhibitory concentration (IC50) of 12 nM. It has a lower affinity to dog liver glucagon receptor with an IC50 of 75 nM. In a recent study, it was shown that NNC 25-0926 significantly inhibits hepatic glucose production (approximately 60%) through its suppressive effect on glycogenolysis but not on gluconeogenesis. Moreover, chronic administration of NNC 25-0926 to high fat diet fed mice showed an improvement in β-cell function, insulin sensitivity and glucose tolerance. Therefore, determining if the chronic usage of NNC 25-0926 in type 2 diabetic patients would improve glycemia is highly recommended. (Rivera et al., 2007).

Type 2 Diabetes Mellitus exhibits a failure in the decrease of Glucagon concentration or even an increase after oral glucose digestion. The lack of Glucagon suppression consequently gives rise to postprandial hyperglycemia in individuals suffering from type 2 diabetes mellitus. The exact cause of this failure remains unclear, but research has shed some light and has indicated that early alpha-cell specific islet defect might be a probable cause. GLP-1 has been shown to inhibit glucagon secretion. In one recent study, the role of glucagon secretion and reduced incretin effect was studied in type 2 diabetic patients. Before the study was performed, all patients with type 2 diabetes had not taken their glucose-lowering agents, if any, for a period of no less than 1 week. Isoglycaemia during 2 days was obtained in both the control group and the group with type 2 diabetes. In the control subjects no difference in glucagon suppression during the first 45 min of OGTT and isoglycaemic i.v. glucose infusion (−36±12 vs −64±23 mmol/l×45 min; p=NS) was observed, whereas in the group of patients with type 2 diabetes significant glucagon suppression only occurred following isoglycaemic i.v. glucose infusion (−63±21 vs 10±16 mmol/l×45 min; p=0.002). Furthermore, in the control group, plasma glucagon levels were shown to be approximately 7.5 pmol/L upon fasting, while diabetic patients exhibited significantly elevated fasting levels of glucagon (10.1 pmol/L). The incretin effect was significantly reduced in patients with type 2 diabetes compared with control subjects, but no significant differences in the secretion of glucagon-like peptide-1 or glucose-dependent insulinotropic polypeptide between the two groups during OGTT or isoglycaemic i.v. glucose infusion, respectively, could explain this. The study had shown that attenuated and delayed glucagon suppression in patients with type 2 diabetes occurs after oral ingestion of glucose, while isoglycaemic i.v. administration of glucose results in normal suppression of glucagon. Intriguingly, a 30% decrease was observed in the suppression of glucagon secretion after oral glucose ingestion compared to intravenous glucose infusion (Knop et al., 2007). It is unclear however, how this mechanism works. We suggest that this phenomenon contributes to both the glucose intolerance and to the reduced incretin effect observed in patients with type 2 diabetes.

The incretin hormone GLP-1 is considered as another natural glucagon antagonist, which inhibits α-cell secretion. In a recent study by De Marinis *et al*, it was shown that GLP-1 suppresses glucagon secretion through PKA dependent and not glucose dependent pathways. Moreover, insulin and somatostatin do not mediate this process. GLP-1 has no known effect on cellular electrical activity, however, it was shown that GLP-1 selectively inhibits N-type Ca^{2+} channels and exocytosis. De Marinis et al, proposed that the PKA-dependent inhibition of N-type Ca^{2+} channels and hence, glucagon secretion by GLP-1 is through a small increase in intracellular cAMP ([cAMP]i). Interestingly, it was concluded that GLP-1 exerts its inhibitory effect on glucagon secretion through activation of a novel GLP-1 receptor undetected by PCR. This conclusion was reached after it was observed that this inhibitory effect was abolished upon the administration of GLP-1 receptor antagonist exendin (9-39) (De Marinis et al., 2010).

Conclusion

In conclusion, antagonists of glucagon receptor are underway to treat diabetes mellitus. In fact, inhibition of alpha cell secretion by the incretin hormone GLP-1 has been proved to be potent and effective. This is true for type 2 diabetic patients and not verified but most probably true as well for type 1 diabetic patients (Hare, 2010). The limitation however of GLP-1 treatment, is its short physiological half-life of 1.5 to 2 minutes, due to its rapid degradation by the enzyme DPP-IV. This renders it as a not so good therapeutic agent for the treatment of diabetes. This is where GLP-1 mimetics, such as Exendin-4 come into play. Exendin-4 is a naturally occurring GLP-1 analogue which was discovered in the saliva of the *Heloderma Suspectum*- the Gila monster (Zhou et al., 2008). This long acting GLP-1 receptor agonist has an N-terminus almost identical to that of the GLP-1, except that the second amino acid residue is glycine in Exendin-4 while alanine in GLP-1 (Hui et al.,

2005) (Figure 3). This one amino acid difference makes it resistant to degradation by the enzyme DPP-IV, hence explaining its long half-life in vivo (Lovshin and Drucker. 2009). Recently, it was shown that the diabetic alpha cell's sensitivity to GLP-1 induced inhibition is preserved. This is a rather unexpected observation, since previous studies have shown the opposite. This phenomenon can be explained as such: alpha cell sensitivity or the extent GLP-1 induced reduction of glucagon secretion is dependent on the patient and the duration of the disease (Hare, 2010). Moreover, treatment of diabetic patients with GLP-1 analogue Exendin-4 or DPP-IV inhibitors show a reduction in glucagon levels, while insulin levels remain the same. This fact suggests that GLP-1 based therapy exhibits a glucagonostatic effect by which the overall blood glucose levels are reduced. Finally, GLP-1 therapy seems to normalize the insulin to glucagon ratio rather than boost absolute insulin levels (Hansen et al., 2011).

References

Alberti K and Zimmet PZ. for the WHO Consultation. Definition, diagnosis and classification of diabetes mellitus and its complications. Part 1: Diagnosis and classification of diabetes mellitus. Provisional report of a WHO consultation. *Diabet.Med.* 15: 7: 539-553, 1998.

Baggio LL and Drucker DJ. Biology of incretins: GLP-1 and GIP. *Gastroenterology* 132: 6: 2131-2157, 2007.

Beauchamp MC, Yasmeen A, Knafo A and Gotlieb WH. Targeting insulin and insulin-like growth factor pathways in epithelial ovarian cancer. *J.Oncol.* 2010: 257058, 2010.

Berghofer P, Peterson RG, Schneider K, Fehmann HC, Goke B. Incretin hormone expression in the gut of diabetic mice and rats. *Metabolism*. 1997 Mar; 46(3):261-267.

Berne R, Levy M, Koeppen B and Stanton B. *The Endocrine System: In Physiology*. 2004, pp. 787-789.

Berthoud HR. Paying the price for eating ice cream: is excessive GLP-1 signaling in the brain the culprit? *Endocrinology* 149: 10: 4765-4767, 2008.

Boron WF and Boulpaep EL. The Endocrine Pancreas. In: *Medical Physiology*, Elsevier Saunders, 2005, p. 1066-1080.

Bowen RA, Austgen l, Rouge M. *Pathophysiology of the endocrine system.* Colorado State University. 2006 Apr.

Bullock BP, Heller RS, Habener JF. Tissue Distribution of Messenger Ribonucleic Acid Encoding the Rat Glucagon-Like Peptide-1 Receptor. *Endocrinology*. 1996; 137(7):2968-2978.

D'Alessio DA, Vahl TP. Glucagon-like peptide 1: evolution of an incretin into a treatment for diabetes. *American Journal of Physiology: Endocrinology and Metabolism*. 2004; 286:882-890.

De Marinis YZ, Salehi A, Ward CE, Zhang Q, Abdulkader F, Bengtsson M, Braha O, Braun M, Ramracheya R, Amisten S, Habib AM, Moritoh Y, Zhang E, Reimann F, Rosengren AH, Shibasaki T, Gribble F, Renstro E, Seino S, Eliasson L and Rorsman P. GLP-1 Inhibits and Adrenaline Stimulates Glucagon Release by Differential Modulation of N- and L-Type Ca2+ Channel-Dependent Exocytosis. *Cell metabolism*. 11, 543–553, 2010.

De Meyts P and Whittaker J. Structural biology of insulin and IGF1 receptors: implications for drug design. *Nat.Rev.Drug Discov*. 1: 10: 769-783, 2002.

Drucker DJ. Enhancing Incretin Action for the Treatment of Type 2 Diabetes. *Diabetes Care*. 2003 Oct; 26(10):2929-2940.

Drucker DJ. The role of gut hormones in glucose homeostasis. *J.Clin.Invest.* 117: 1: 24-32, 2007.

Eknoyan G and Nagy J. A history of diabetes mellitus or how a disease of the kidneys evolved into a kidney disease. *Adv.Chronic Kidney Dis.* 12: 2: 223-229, 2005.

Fehmann HC, Goke R, Goke B. Glucagon-like peptide-1(7-37)/(7-36)amide is a new incretin. *Molecular and Cellular Endocrinology*. 1992 May; 85(1-2):C39-44.

Gautier JF, Fetita S, Sobngwi E, Martin CS. Biological actions of the incretins GIP and GLP-1 and therapeutic perspectives in patients with type 2 diabetes. *Diabetes Metabolism*. 2005; 31:233-242.

Gomez E, Pritchard C, Herbert TP. cAMP-dependent Protein kinase and Ca2+ Influx through L-type Voltage-gated Calcium Mediate Raf-independent Activation of Extracellular Regulated Kinase in Response to Glucagon-like Peptide-1 in Pancreatic β-Cells. *The Journal of Biological Chemistry*. 2002 Sep; 277(50):48146-48151.

Gros R, You X, Baggio LL, Kabir MG, Sadi A, Mungrue IN, Parker TG, Huang Q, Drucker DJ, Husain M. Cardiac Function in Mice Lacking Glucagon-Like Peptide-1 Receptor. *Endocrinology*. 144(6):2242-2252, 2006.

Hansen M, Hare KJ, Holst JJ and Knop FK. Inhibition of Glucagon Secretion by GLP-1 Agonists and DPP4 Inhibitors. *Journal of Clinical Metabolism and Diabetes.* 2:(1) 2011.

Hare KJ. Role of GLP-1 induced glucagon suppression in type 2 diabetes mellitus. *Danish Medical Bulletin.* 57: [B4181], 2010.

Hinnen D, Loretta L, Waninger A, Kushner P. Incretin Mimetics and DPP-IV Inhibitors: New Paradigms for Treatment of Type 2 Diabetes. *Journal of the American Board of Family Medicine.* 2006; 19(6):612-620.

Holst J. *Glucagon-like peptide 1: from extract to agent.* The Claude Bernard Lecture, 2005. Diabetologia. 2005.

Holt RI. Diagnosis, epidemiology and pathogenesis of diabetes mellitus: an update for psychiatrists. *Br.J.Psychiatry Suppl.* 47: S55-63, 2004.

Hui H, Zhao X and Perfetti R. Structure and function studies of glucagon-like peptide-1 (GLP-1): the designing of a novel pharmacological agent for the treatment of diabetes. *Diabetes Metab.Res.Rev.* 21: 4: 313-331, 2005.

Inzucchi SE and McGuire DK. New drugs for the treatment of diabetes: part II: Incretin-based therapy and beyond. *Circulation* 117: 4: 574-584, 2008.

Kinzig KP, D'Alessio DA and Seeley RJ. The diverse roles of specific GLP-1 receptors in the control of food intake and the response to visceral illness. *J.Neurosci.* 22: 23: 10470-10476, 2002.

Knop FK, Vilsbøll T, Madsbad S, Holst JJ and Krarup T. Inappropriate suppression of glucagon during OGTT but not during isoglycaemic i.v. glucose infusion contributes to the reduced incretin effect in type 2 diabetes mellitus. *Diabetologia.* 50:797–805, 2007.

Lovshin J and Drucker DJ. Synthesis, secretion and biological actions of the glucagon-like peptides. *Pediatric Diabetes.* 2000; 1:49-57.

Lugari R, Dei Cas A, Ugolotti D, Finardi L, Barilli AL, Ognibence C, Luciani A, Zandomeneghi R, Gnudi A. Evendence for early impairment of glucagon-like peptide 1 inducing insulin secretion in human type 2 (non insulin dependent) diabetes. *Horm Metab Res.* 2002 Mar; 34(3):150-154.

Mattson MP, Perry TA, Greig NH. Learning from the gut: An intestinal peptide that regulates glucose metabolism by stimulating insulin secretion from the pancreas is now shown to enhance learning and memory. The peptide can also protect nerve cells in models of neurodegenerative disorders. *Nature Medicine.* 2003 Sep; 9(9):11731179.

Mojsov. Glucagon-like Peptide-1 (GLP-1) and the Control of Glucose Metabolism in Mammals and Teleost Fish. *Amer Zool.* 2000; 40:246-258.

Noyan T, Balaharoğlu R and Kömüroğlu U. The oxidant and antioxidant effects of 25-hydroxyvitamin D3 in liver, kidney and heart tissues of diabetic rats. *Clin Exp Med*.5: 1: 31-36, 2005.

Parker JC, McPherson RK, Andrews KM, Levy CB, Dubins JS, Chin JE, Perry PV, Hulin B, Perry DA, Inagaki T, Dekker KA, Tachikawa K, Sugie Y, Treadway JL.Pessin JE and Saltiel AR. Signaling pathways in insulin action: molecular targets of insulin resistance. *J.Clin.Invest.* 106: 2: 165-169, 2000.

Poretsky, l. *Principles of Diabetes Mellitus*. London: Kluwer Acedemic Publishers; 2002.

Rhoades RA and Bell DR. The Endocrine Pancreas. In: *Principles for Clinical Medicine*. 3rd Edition. Lippincott Williams and Wilkins, p.641-655, 2009.

Rivera N, Everett-Grueter CA, Edgerton DS, Rodewald T, Neal DW, Nishimura E, Larsen MO, Jacobsen LO, Kristensen K, Brand CL and Cherrington AD. A Novel Glucagon Receptor Antagonist, NNC 25-0926, Blunts Hepatic Glucose Production in the Conscious Dog. The *Journal of Pharmacology and Experimental Therapeutics*, 321:743–752, 2007.

Robles GI and Singh-Franco D. A review of exenatide as adjunctive therapy in patients with type 2 diabetes. *Drug Des.Devel.Ther.* 3: 219-240, 2009.

Seifter J, Ratner A, Sloane D. The Endorine Pancreas: Fed and Fasted Metabolic States. In: *Concepts in Medical Physiology*. Lippincott Williams and Wilkins:.498-503, 2005.

Suzuki A, Nakauchi H, Taniguchi H. Glucagon-like peptide 1(1-37) converts intestinal epithelial cells into insulin-producing cells. *Proceedings of the National Academy of Sciences of the United States of America*. 2003 April; 100(9):50345039.

Wei Y, Mojsov S. *Tissue-specific expression of the human receptor for glucagon-like peptide-I: brain, heart and pancreatic forms have the same deduced amino acid sequences.* Federation of European Biochemical Societies. 1995; 358:219-224.

Yusta B, Baggio LL, Estall JL, Koehler JA, Holland DP, Li H, Pipeleers D, Ling Z, Drucker DJ. GLP-1 receptor activation improves beta cell function and survival following induction of endoplasmic reticulum stress. *Cellular Metabolism*. 2006 Nov; 4(5):391406.

Zhou J, Chu J, Wang YH, Wang H, Zhuang YP and Zhang SL. Purification and bioactivity of exendin-4, a peptide analogue of GLP-1, expressed in Pichia pastoris. *Biotechnol.Lett.* 30: 4: 651-656, 2008.

In: Glucagon
Editor: Ahmed Al-Jebawi
ISBN: 978-1-62618-147-2

Chapter 6

Glucagon Antagonists

Alaaeldin M. K. Bashier* and Puja Murli Thadani
Endocrine division-Medical department,
Dubai Hospital, United Arab Emirates

Glucagon Receptor Antagonists

Abstract

Despite the fact that glucagon was discovered two years after the discovery of insulin, little data is available on the physiology of the alpha cells. This could be attributed to the scarcity of this cell population in animal models such as mice and rats and the fact that functional identification methods have been lacking. In the recent years, much progress has been made in the study of alpha cell function, physiological role of glucagon and glucagon receptor as well as the intracellular pathways and messengers. Furthermore, there are many studies that proved the relationship between hyperglucagonemia and diabetes mellitus. The available data on this association allowed researchers to target glucagon as a valid therapeutic option to treat hyperglycemia in patients with diabetes mellitus.

* Correspondence: Dr. A.M Bashier. Email. Alaaeldin11@gmail.com, P.O.Box 94132. Dubai. UAE.

In this chapter we discussed the physiological steps in the synthesis and secretion of glucagon, the relation between glucagon and diabetes as well as the new molecules targeting glucagon and glucagon receptors.

Introduction

Pancreatic cell function is totally dependent on co-ordinated secretion of beta and alpha cells. These cells respond to changes in blood glucose in a totally opposite way, for example beta cell secretion of insulin is known to increase in case of high blood glucose levels, while alpha cell secretion of glucagon increases in case of hypoglycaemia (Nadal et.al. 1999, Quesada et.al. 2006). It is known that insulin, being an anabolic hormone, causes accumulation of glycogen and fat, hence reduction in blood glucose levels. On the other hand, glucagon induces a catabolic state, by stimulating hepatic glycogenolysis, and gluconeogenesis, which results in an increase in blood glucose.

Several decades of research and evolving technology have led to the isolation, purification and sequencing of the 29-amino acid peptide hormone, Glucagon. Additionally, the structure of the proglucagon gene has been identified and its tissue-specific processing has been established over the years. Proglucagon is processed into functional glucagon by subtilisin-like proprotein convertase-2 (PC2) in the pancreatic alpha-cells (Rouille et al., 1994). Glucagon acts via a seven transmembrane G-protein coupled receptor consisting of 485 amino acids (Jelinek et al., 1993). The binding sites for this peptide hormone are present in multiple tissues including liver, brain, pancreas, kidney, intestines and adipose tissue (Burcelin et al., 1996 and Christophe, 1996). Such a detailed understanding of this peptide hormone has allowed researchers to target glucagon as a valid therapeutic option in patients with diabetes mellitus.

Structure of Islet Cells

The pancreatic islet is a multicellular structure composed of beta, alpha, delta, and pancreatic polypeptide cells. In a healthy human adult, the pancreas contains almost 1 million islets that would represent almost 1-2 % of the total pancreatic mass. Each islet is made up of 1000-3000 cells. In rats, beta cells constitute 60–80 %, alpha cells 15–20 %, and delta cells < 1% (Brelje et.al

1998, Brissova et.al 2005). In humans, the distribution of islet cells is somewhat different with equal numbers of delta and pancreatic polypeptide cells, and a smaller beta cell mass, which represent 48-59 % of total number of cells. The alpha cells contribute to the remaining 33 -46% of the islet cell mass. This higher proportion of alpha cell mass would suggest a more important role for glucagon in humans (Cabrera et.al 2006).

There are many structural differences between rodent and human islets. In rodent islets, the clustering of beta cells makes their responses to elevated glucose concentrations more coordinated with a synchronous oscillatory response and a typical pulsatility in insulin release (Vozzi et. al 1995). On the other hand, due to lack of clustering of human beta cells, these oscillations are not synchronized throughout the human islets. Although this unique cytoarchitecture of human beta cells makes their activity discontinuous, it allows them to intermingle freely with other endocrine cells throughout the islet with a more efficient paracrine connection between different cell types.

The high vascularity of the pancreatic islets helps pancreatic endocrine cells to sense variations in plasma glucose more accurately and hence, allows appropriate secretory response. Ahren (2000) suggested a multi-level regulation of pancreatic islets, which involves parasympathetic, sympathetic & sensory nerves.

Synthesis of Glucagon

The proglucagon gene, which is expressed in the central nervous system, entero-endocrine cells (L-cells of the intestines) and the islet alpha cells encodes for glucagon, GLP-1, and GLP-2. Different prohormone convertase enzymes are responsible for the generation of these peptides (mojsov et al. 1986). In the alpha cells the predominant enzyme is PCSK2 that leads to production of glucagon, Glicentin, Glicentin-related peptide, intervening peptide 1, and major proglucagon fragment (Dey et al. 2005). In the intestines the predominant enzyme is the PCSK1/3 that results in production of GLP-1, GLP-2, oxyntomodulin, Glicentin, and intervening peptide 2 (mojsov et al. 1986). The absence of PCSK2 in knockout-mice results in failure of mature glucagon production (Furuta et al. 2001).

There are many factors that affect glucagon gene expression. Glucagon gene expression is negatively regulated by insulin (Philippe et al. 1995). In diabetic rats and in conditions of insulin deficiency and hyperglycemia, glucagon gene expression would be augmented and correction of insulin levels would be associated with normalization of plasma glucose and glucagon levels (Dumonteil et al. 1998). The mechanism of insulin inhibition of glucagon secretion involves activation of GABA receptor translocation to cell surface in an Akt-dependent manner (Basu A et al. 2004).

The amino acid arginine increases glucagon gene expression in rats; this process is mediated by protein kinase C. On the other hand, other nutrients like fatty acids may play different roles in gene expression; an example is palmitate, which down regulates glucagon gene expression (Dumonteil et al. 2000). Zinc secretion from beta cells is another important factor in glucagon secretion inhibition. Impaired zinc secretion results in augmented glucagon secretion in response to hypoglycemia (Zhou H et al. 2007).

In conclusion, insulin suppresses glucagon gene expression and therefore, conditions of insulin deficiency may cause hyperglucagonemia. While hyperglycemia stimulates beta cell insulin secretion, it inhibits alpha cell secretion of glucagon. Furthermore, Mac Donald et al., (2007) and Vieira et al., (2007) have proved that glucagon inhibition occurs at glucose levels below the threshold of beta cell activation.

Glucagon Secretion

Alpha cells are equipped with multiple channels that regulate glucagon secretion. As this process involves many intracellular steps, blockage at any level would result in impaired glucagon secretion.

KATP channels play a fundamental role in the process of glucagon secretion. This is similar to the role they play in beta cells with one important difference in the sensitivity to ATP. As alpha cell KATP has a lower sensitivity to ATP (Bokvist et al. 1999, Gromada et al. 2007), maximal inhibition of the KATP channels requires a lower ATP concentration.

When the K_{ATP} channels are activated, cell membrane depolarization with an action potential of -60Mv results in activation of T-type calcium channels. They act as a pacemaker of the action potential (Gopel etal, 2000). At an action potential of -30 to -20mV, Na+ channels and L-type Ca++ channels are activated. Although most of current calcium goes through L-type Ca++ channels, calcium required for exocytosis of glucagon comes mainly through N-type Ca++ channels (Gromada et al. 1997, 2004, Olsen et al. 2005). (See figure 1)

The voltage dependent K+-channels mediate the membrane repolarization. Also the increase in extracellular glucose raises the cytosolic ATP/ADP ratio that blocks the K_{ATP} channels, inactivating membrane depolarization.

The sodium channels are tertradotoxin sensitive and their blockage by tertradotoxin inhibits glucagon secretion (Gromada et al, 2004).

Glucagon Receptor

Glucagon exerts its action on target tissues through activation of the glucagon receptor, a G-protein coupled receptor member of the class II G-protein-coupled receptor superfamily (Jelinik et al., 1993). Activation of the glucagon receptor leads to signal transduction by G proteins (Gsa and Gq).

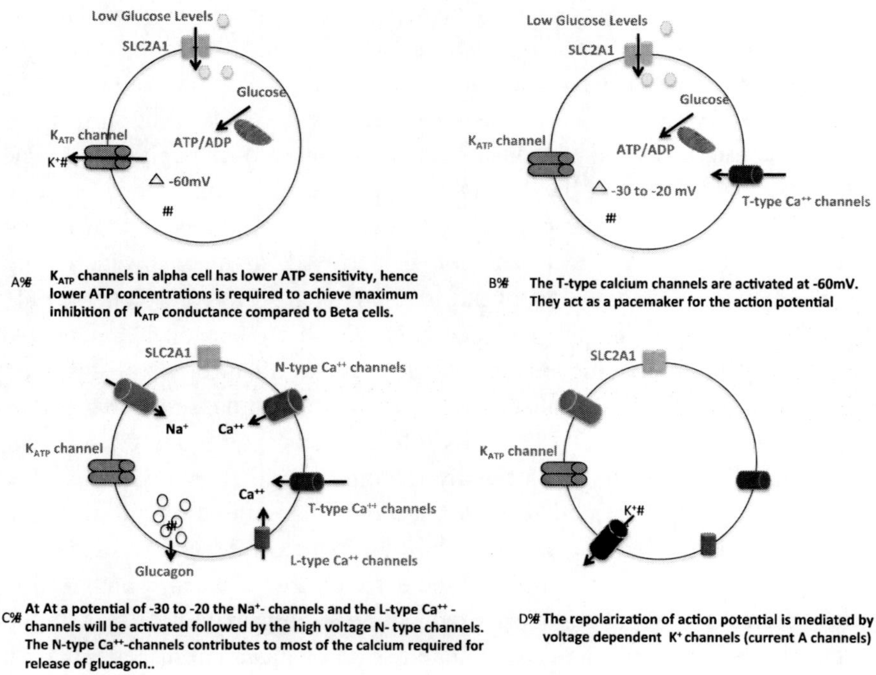

Figure 1. A stepwise Schematic module for the glucose dependent glucagon secretion in alpha cell.

Gsa activates adenylate cyclase, which causes cAMP production, resulting in an increase in levels of protein kinase, whereas Gq activation leads to phospholipase-C mediated increases in intracellular calcium. Glucagon receptor signalling in the liver results in increased hepatic glucose production by stimulation of glycogenolysis and gluconeogenesis in addition to inhibition of glycogenesis. These receptors are also expressed in extra-hepatic tissues, including the heart, intestinal smooth muscle, kidney and brain, however, much less is known about the action of glucagon in these tissues (Hansen et al., 1995).

Physiological Role of Glucagon

In parallel to investigating the biochemical properties of this pancreatic peptide hormone, the physiological actions of glucagon have been explored. The administration of exogenous glucagon was found to increase glucose levels in fasted and fed animals (Myers et al., 1991 and Young et al., 1993), and similar observations were noted in humans (Lins et al., 1983 and Hvidberg et al., 1994). Freychet et al., (1988) also demonstrated that glucagon raises plasma glucose levels in response to insulin induced hypoglycemia. The current use of glucagon in the management of severe hypoglycemia is based on this observation. Moreover, in vivo studies have directly shown that glucagon stimulates glucose production in perfused rat livers by increasing both glycogenolysis and gluconeogenesis (Ikeda et al., 1989 and Beuers et al., 1990). A substantial amount of evidence proves that glucagon is a sensitive and timely regulator of glucose metabolism in vivo. Small doses of glucagon are sufficient to induce significant elevations in blood glucose levels (Lins et al., 1983 and Myers et al., 1991). This effect of glucagon is known to occur within minutes and it dissipates rapidly (Dobbins et al., 1998). On the other hand, there is ample evidence to demonstrate that inhibition of glucagon signalling in vivo leads to a reduction in blood glucose levels. A significant drop in glucagon levels was noted in pancreatectomized canines, which lead to marked decrease in hepatic glucose production (Stevenson et al., 1987). In the subsequent years, it was shown that suppression of glucagon in patients with diabetes improved glucose tolerance (Dineen et al., 1995 and Shah et al., 2000). These studies highlighted the strong association between elevated glucagon levels and hyperglycemia. In summary, glucagon is secreted into the bloodstream from the islets in a pulsatile fashion (Opara et al., 1988) in response to low blood glucose levels. It is the pulsatile delivery of glucagon that is more effective in inducing hepatic glucose production in vitro (Komjati et al., 1986), ex vivo (Paolisso et al., 1989) and in vivo (Weigle et al., 1986). Hence, the main physiological role of glucagon is to stimulate hepatic glucose output, thereby leading to an increase in glucose levels. This provides the major counter-regulatory mechanism for insulin in maintaining glucose homeostasis.

Effect on Glycogenolysis

Glucagon secretion is essential for providing the body with energy by increasing the breakdown of glycogen and the simultaneous inhibition of hepatic glycogen synthesis. Glucagon activates Protein kinase-A (PKA), which through a cascade of phosphorylation processes results in phosphorylation of glycogen and ultimately results in increased glycogen breakdown, i.e., glycogenolysis and increased Glucose-6-phosphate (G-6-P) production. G-6-P, a precursor of glucose, will then be converted to glucose by glucose-6-phosphatase, hence increasing the hepatic glucose output (Krebs 1981 and Johnson et al., 1997).

Effect on Glycogenesis

In addition to promoting glycogenolysis, glucagon inhibits glycogen synthesis in the liver by regulating glycogen synthase. This enzyme plays a key role in glycogen synthesis as it catalyses the transfer of glucosyl residue from UDP-glucose to a non-reducing end of the branched glycogen molecule. Glucagon induces glycogen synthase phosphorylation and inhibits glycogen synthase activity in the liver (Ramachandran et al., 1983). Inactivation of glycogen synthase reduces the synthesis of glycogen (Roach et al., 1990). Therefore, the role of glucagon is to inhibit glycogenesis by inactivating glycogen synthase, consequently increasing the hepatic glucose output.

Effect on Gluconeogenesis

In addition to its effect on glycogen metabolism, glucagon regulates blood glucose levels by increasing gluconeogenesis. Phosphoenolpyruvate carboxykinase (PEPCK) catalyses the conversion of oxaloacetate to phosphoenolpyruvate and, is an early and rate limiting step in the pathway of hepatic gluconeogenesis. Earlier studies have shown that treatment with glucagon increases the levels of PEPCK mRNA in hepatocytes (Beale et al., 1984 and Christ et al., 1988). More recent data suggests that PKA activation by cAMP leads to phosphorylation of cAMP response element binding (CREB) protein. The phosphorylated CREB protein in turn binds to a cAMP-responsive element and up-regulates peroxisome proliferator activated receptor- co-activator-1 (PGC-1) (Herzig et al., 2001). Consequently, PGC-1 and the nuclear transcription factor hepatocyte nuclear factor-4 (HNF-4) act together to increase transcription of the PEPCK gene (Yoon et al., 2001). Given that glucagon activates PKA, it is likely that glucagon-mediated up-regulation of PEPCK transcription and activity leads to increased gluconeogenesis in the liver.

Effect on Glycolysis

In addition to increased gluconeogenesis, glucagon inhibits glycolysis. The conversion of fructose-6-phosphate into fructose-1,6 bisphosphate by Phosphofructokinase-1 (PFK-1) is an early and important rate limiting step in glycolysis. The rate limiting enzyme, PFK-1, is allosterically activated by fructose-2,6 bisphosphate (Okar et al., 1999). Glucagon reduces the levels of fructose-2, 6 bisphosphate resulting in inhibition of PFK-1, thereby inhibiting glycolysis (Veneziale et al., 1976 and Castano et al., 1979).

Pyruvate kinase catalyses the transfer of the phosphate group from phosphoenolpyruvate to ADP resulting in the production of pyruvate and ATP. This is the final step in the glycolysis pathway. Glucagon inhibits pyruvate kinase by several mechanisms. It also inhibits transcription of the pyruvate kinase gene and increases the degradation of pyruvate kinase mRNA (Pilkis et al., 1991). Hence, glucagon inhibits pyruvate kinase, which results in decreased glycolysis and increased gluconeogenesis.

Relationship between Glucagon, the Glucagon Receptor and Diabetes

Although glucagon was discovered in 1923 it had not been linked to diabetes until 1975. In that year R. Unger published his new theory that changed the way scientists looked at diabetes management. The "uni-hormonal theory", that originated with the work of von Mering and Minkowski in 1981 and was reinforced by those of Banting and Best in 1922, states that all the metabolic derangement in diabetes mellitus are attributed to absolute or relative insulin deficiency. The new R. Unger "Bi-hormonal theory" states that diabetic abnormalities in glucose homeostasis are the consequences of a bi-hormonal disorder in which a relative or absolute deficiency of insulin and a relative or absolute excess of glucagon are etiological factors. This theory indicates that hypo-secretion of insulin would account primarily for underutilization, and relative hyper-secretion of glucagon leads to over production of glucose (Unger et al. 1975). The hypothesis was based on the fact that somatostatin induced inhibition of glucagon and insulin does not result in hyperglycemia unless glucagon deficiency is corrected or restored by exogenous glucagon. Moreover somatostatin reduction of glucagon was found to be associated with reduction in blood glucose levels. Another important evidence is that hyperglucagonemia has been seen with all conditions associated with endogenous hyperglycemia including spontaneous diabetes in humans, and diabetes in experimental animals, and hyperglycemia associated with burns, myocardial infarction, etc. (Unger et al. 1975).

Further evidence on the contribution of glucagon to hyperglycemia was shown by Felig et al., (1976), who studied the effect of glucagon infusion in overnight fasted healthy subjects on arterial glucagon, insulin and glucose levels, and on splanchnic glucose output. During the infusion of glucagon, plasma glucagon levels rose rapidly, reached a plateau within 15 minutes, and remained elevated for 45 minutes. The arterial glucose and insulin concentrations also increased rapidly and reached a maximal level by 30 minutes. Despite the concurrent increase in plasma insulin levels, splanchnic glucose output rose during the glucagon infusion. The glucose output was three times more than the basal levels within 7.5 minutes, remained elevated for approximately 15 minutes and then fell progressively.

Importantly, this study established that high physiological levels of glucagon increase glucose production despite a parallel increase in plasma insulin levels. In a second experiment, Felig et al., (1976) infused glucose at a rate that was sufficient to completely suppress glucose production prior to infusing glucagon. Under these conditions, the glucagon infusion reversed the glucose and insulin mediated suppression of splanchnic glucose production. It is noteworthy that in these experiments on healthy subjects the glucagon-stimulated increase of glucose production was transient.

We know now that type2 diabetes is associated with insulin deficiency and a state of hyperglucagonemia. This circulating glucagon or an increase in glucagon/insulin ratio is an important determinant of fasting and postprandial hyperglycaemia seen in type2 diabetes patients (Baron et al. 1987, Basu et al. 2004, Gastaldelli et al. 2000).

In patients with Type 2 Diabetes, the degree of fasting hyperglycaemia correlates with the rate of basal hepatic glucose output suggesting that the extent of this defect may be an important determinant of the severity of diabetes (Baron et al., 1987 and DeFronzo et al., 1989). The increase in hepatic glucose production in patients with overt Type 2 Diabetes is primarily due to increased gluconeogenesis (Consoli et al., 1989 and Magnusson et al., 1992). However, impaired insulin mediated suppression of glucose production has also been implicated in patients with 'mild' hyperglycemia (Basu et al., 2004).

There is recent data that shows the benefit of reducing glucagon action in glucagon receptor knockout ($Gcgr^{-/-}$) mice (Lee et al. 2011). The study of Lee et al. aimed to determine if glucagon causes the lethal consequences of insulin deficiency. Surprisingly, $Gcgr^{-/-}$ knockout mice appeared to be healthy and were completely protected from diabetes compared to streptozocin treated $Gcgr^{+/+}$ mice, which became severely hyperglycaemic (Lee et al. 2011). $Gcgr^{-/-}$ knockout mice had normal fasting blood glucose, and astonishingly, despite the absence of a rise in plasma insulin, they demonstrated a normal or even an improved response to glucose tolerance test. The authors concluded that insulin action is largely directed towards overcoming the hepatic actions of glucagon. This proves the strong evidence supporting a role for glucagon in contributing to diabetic hyperglycemia (Dale 2011).

Relation to Fasting Hyperglycaemia

Early evidence supporting the theory that glucagon contributes to the development of fasting hyperglycemia was provided by a study on patients with Type 1 Diabetes, in whom an infusion of insulin was used to maintain euglycemia (Gerich et al., 1975). On discontinuation of insulin, these patients were given an infusion of saline or somatostatin. Gerich et al. had shown that glucagon suppression by somatostatin prevents the development of diabetic ketoacidosis and significantly improves fasting hyperglycemia during the phase of insulin withdrawal. The authors of this study concluded that hypoinsulinemia by itself does not lead to fasting hyperglycemia and ketoacidosis, and that glucagon through its glycogenolytic, gluconeogenic, ketogenic and lipolytic effects contributes to the full development of this condition.

In a following study, Gerich *et al.*, (1976a) further assessed the contribution of glucagon towards the development and maintenance of fasting hyperglycemia. Patients with Type 1 Diabetes were given an overnight insulin infusion, which was then terminated and replaced by saline infusion for 6 hours to allow development of marked hyperglycemia and ketoacidosis. At this time, saline infusion was either continued or replaced by somatostatin infusion for an additional 6 hours. Among patients who were given somatostatin infusion, the plasma glucose levels fell progressively in the next 6 hours. Also, the levels of plasma β–hydroxybutyrate declined significantly in patients receiving somatostatin infusion compared to those on saline infusion. The results from this study demonstrated rapid and marked suppression of plasma glucagon levels with somatostatin infusion, and concluded that during hypoinsulinemia, increased endogenous glucagon secretion significantly contributes towards fasting hyperglycemia. This conclusion was reinforced by another study, which reported that replacement of glucagon reversed the glucose lowering effects of somatostatin infusion (Gerich *et al.*, 1976b).

Moreover, similar findings have been observed in studies including patients with Type 2 Diabetes where somatostatin was used with or without insulin replacement. Baron et al., (1987) demonstrated that infusion of somatostatin, without insulin replacement, reduced hepatic glucose production by approximately 25%. On the other hand, infusion of somatostatin with basal insulin replacement reduced hepatic glucose production by 58%. Hence, these studies provided evidence that elevated glucagon levels significantly contribute to increased hepatic glucose production in patients with Type 2 Diabetes; thereby contributing to significantly elevated fasting blood glucose levels.

In a study where glucagon was infused into normal subjects as well as patients with Type 1 and Type 2 Diabetes, the infusion of glucagon led to deterioration of glycemic control in patients with type 1 diabetes who were insulin withdrawn (Sherwin *et al.*, 1976). These findings suggest the primary role of insulin deficiency in the diabetogenic action of glucagon. Due to the fact that normal β–cells are able to increase their insulin secretion to counteract the action of glucagon, hyperglucagonemia does not cause glucose intolerance in normal subjects. Interestingly, the three poorly controlled patients with Type 2 Diabetes included in the study did not experience further deterioration in their glycemic control probably because they were already responding maximally to their inappropriately elevated endogenous level of glucagon. Investigators of this study also concluded that hyperglucagonemia does not cause impaired glucose tolerance in normal subjects or lead to deterioration of glycemic control when endogenous insulin is available in patients with Type 2 Diabetes. The latter conclusion is debatable as the changes in glucose levels may depend on the amount of insulin available. An additional study by Gerich *et al.*, (1977) showed that suppression of glucagon secretion with 3-day infusion of somatostatin in insulin treated patients resulted in markedly improved glycemic control in addition to a 50% reduction of insulin dose in some cases. Collectively, these studies show that it is difficult to accurately ascertain the contributions of α–cell *versus* β–cell dysfunction towards the development and maintenance of fasting hyperglycemia. However, it can be concluded that due to the influence of endogenous insulin on glucagon secretion, defects in β–cell function are invariably accompanied by inappropriately elevated plasma glucagon levels, which contribute to fasting hyperglycemia in patients with diabetes.

Relation to Postprandial Hyperglycemia

It is known that there is an abnormally elevated glucagon levels despite hyperglycemia in both type1 and type2 diabetes. In addition to that, patients with type1 diabetes have diminished or absent glucagon response to hypoglycemia (Clarke et al. 1991). The inappropriately elevated glucagon during hyperglycemia in Type1 diabetes mellitus has been considered to be due to lack of intra-islet insulin to restrain the glucose effect on the function of alpha-cell (Muller et al. 1971).

Gerich et al. conducted one of the earliest studies that documented the role of glucagon in the pathogenesis of postprandial hyperglycemia in 1975. They studied plasma glucose and glucagon responses to normal meals in 12 subjects with type1 diabetes, and 12 normal subjects. They noticed that the suppression of glucagon, that occurs after a carbohydrate meal, fails to show in normal subjects after ingestion of a balanced meal containing carbohydrates, protein, and fats. This glucagon rise peaks at 120min, and is related to the protein content in the meal. Among subjects with type 1 diabetes, glucagon response to a balanced meal was found to be three times higher than that found in normal subjects, and in contrast to normal subjects, it peaked at 30minutes. To evaluate the contribution of glucagon to post prandial glucose levels, Gerich and co-workers found out that somatostatin administration diminishes postprandial hyperglycemia by 60%. Because somatostatin has no direct action on hepatic glucose production or glucose utilization, the reduction in postprandial glucose was attributed to inhibition of glucagon (Gerich et al. 1975). Evidence on the relation between high glucagon levels and type 2 diabetes was shown by Mitrakou et al, who conducted two studies to compare pancreatic hormone levels, and glucose production and utilization by using dual tracer technique during an oral glucose tolerance test (OGTT) in patients with type2 diabetes or impaired glucose tolerance (IGT), and normal subjects. In the first study, they have shown that patients with type2 diabetes, compared to normal subjects, had impaired early insulin release, elevated fasting glucagon levels and impaired glucagon suppression with hyperglycemia. Obviously, these changes contribute to elevation of both fasting and postprandial glucose levels. Furthermore, they demonstrated normal muscle glucose uptake in both groups. They concluded that inadequate suppression of plasma glucagon is primarily responsible for the postprandial hyperglycemia (Mitrakou et al. 1990). In their second study, the comparison was between subjects with impaired glucose tolerance and normal subjects. Postprandial glucagon levels were found to be elevated in subject with impaired glucose tolerance, a finding similar to that shown in patients with type2 diabetes. The only exception was the absence of fasting hyperglycemia in IGT subjects (Mitrakou et al. 1992). Shah et al. tested the hypothesis that a lack of suppression of glucagon causes postprandial hyperglycemia in subjects with type2 diabetes. Nine diabetic subjects ingested 50 g glucose while they were receiving both somatostatin and insulin infusion. In one occasion, patients received glucagon infusion and on the other day, glucagon infusion was delayed by two hours. They concluded that lack of suppression of glucagon levels contributes to postprandial hyperglycemia in subjects with type2 diabetes (Shah et al, 2000).

Therapeutic Approaches Targeting Glucagon Pathways for the Management of Diabetes

Introduction

Parker *et al.*, (2003) demonstrated that targeted disruption of the glucagon receptor gene (GR -/-) in mice led to significant reductions in fasting and fed plasma glucose levels compared to their non-transgenic species littermates (GR+/+) or wild-type mice. Although GR-/- mice had improved glucose tolerance, it is noteworthy that no overt hypoglycemia occurred in this species. In comparison to the control mice, plasma glucagon was markedly elevated to supra-physiological levels under both fasting and fed conditions in GR-/- knockout mice. Furthermore there was a significantly greater pancreatic glucagon content. The investigators concluded that the GR-/- mice were able to maintain near-normal plasma glucose levels in the complete absence of glucagon receptors.

Subsequently, Gelling *et al.*, (2003) carried out a similar investigation in which they evaluated a different strain of GR-/- knockout mice. They reported a reduction in fasting and fed glucose levels along with markedly elevated plasma glucagon levels in GR-/- mice compared to the control group. The GR-/- mice exhibited improved glucose tolerance and increased insulin sensitivity during an insulin tolerance test, but their serum insulin levels remained unchanged in comparison to the control group. Interestingly, the male GR-/- mice experienced frank hypoglycemia during the insulin tolerance test in addition to severe hypoglycemia, defined as blood glucose 1.7mml/l, during a 24-hour fast. This study showed that, in the absence of functional glucagon receptors, mice have significantly decreased glucose levels throughout the day, increased insulin sensitivity and may develop overt hypoglycemia. These findings are consistent with the concept that glucagon plays a physiological role in determining both fasting and post-prandial glucose levels.

Chen *et al.*, (2005) used mice with liver specific ablation of glucagon receptors (LGsKO) in their study. In these rodent models, serum glucose and insulin levels were both significantly decreased in fed LGsKO mice compared to normal littermates. LGsKO mice showed increased glucose tolerance with both increased glucose-stimulated insulin secretion and increased insulin sensitivity in the liver and muscle. Also, LGsKO mice had very high serum glucagon and glucagon-like peptide-1 (GLP-1) levels in addition to pancreatic α–cell hyperplasia, which could either be, secondary to hepatic glucagon resistance or chronic hypoglycemia, or both.

Glucagon Receptor Anti-Sense

Liang *et al.*, (2004) used specific glucagon receptor antisense oligonucleotide (GR-ASO) to understand the anti-diabetic effects of reduced glucagon receptor expression in db/db mice. The ability of GR-ASOs to inhibit glucagon receptor mRNA expression was demonstrated by quantitative (real-time) RT-PCR. GR-ASO was administered intra-peritoneally at a dose of 25mg/kg twice per week for 3 weeks. This study described several noteworthy findings. Firstly, administration of GR-ASOs significantly decreased glucagon receptor mRNA expression in the liver by ~80%, consequently decreasing glucagon stimulated cAMP production by ~47%. These observations are important as after binding to its receptor, glucagon activates adenylyl cyclase, thus triggering glycogenolysis via a cAMP-related signalling pathway. Secondly, there was a significant reduction in levels of blood glucose (~24%), triglycerides (~62%) and free fatty acids (~36%). Thirdly, there was a significant improvement in glucose tolerance (~20%), and lastly, a diminished hyperglycemic effect was noted in response to glucagon challenge. No hypoglycemia episodes were reported in db/db mice. These findings led to the conclusion that partial reduction of glucagon receptor expression ameliorates hyperglycemia in diabetic mice.

Similarly, Sloop *et al.*, (2004) carried out a study to determine the extent to which blocking glucagon action would reverse hyperglycemia using another GR-ASO to target glucagon receptors in diabetic rodent models. In this study, administration of GR-ASOs resulted in reduced plasma glucose levels, α–cell hypertrophy, hyperplasia and reversible hyperglucagonemia, in addition to an increase in pancreatic α–cell expression of active GLP-1. As a result, there was an increase in islet insulin content with preservation of insulin secretion and improved glucose tolerance. GR-ASO therapy also decreased glycogenolytic and gluconeogenic enzyme gene expression, thus preventing glucagon-mediated hepatic glucose production without causing hypoglycemia.

Both studies discussed here clearly show that GR-ASO treatment improved glycemic control through its hepatic and islet effects in rodent models of diabetes. These results are consistent with previous studies testing the use of anti-glucagon molecules, where transient blockade of glucagon action in the liver led to a significant reduction in hepatic glucose production. The data from these studies provided early evidence suggesting that such molecules might be useful in the treatment of post-prandial hyperglycemia, and could potentially be used to treat impaired glucose tolerance and Type 2 Diabetes. Due to the crucial role that glucagon plays in stimulating hepatic glucose output, other possible effects of targeting the glucagon receptor include hypoglycemia and excessive accumulation of glycogen in the liver. It is important to note that neither hypoglycemia nor any adverse hepatic effects have been reported with GR-ASO therapy in these studies.

Neutralizing Glucagon Antibodies

Studies using a high-affinity, neutralizing monoclonal antibody provide stronger evidence for the role of glucagon in determining fasting and post-prandial glucose levels in addition to its contribution to hyperglycemia in diabetic animal models. Monoclonal antibodies to glucagon which have been shown to completely block the hyperglycemic effect of exogenous glucagon infusion in normal rats were shown to reduce blood glucose levels by ~44% in moderately diabetic STZ-induced rats. Comparatively, in severely diabetic STZ-induced rats, immunoneutralization of glucagon with monoclonal antibodies was ineffective. Interestingly, chronic insulin treatment in these rats restored the effectiveness of glucagon immunoneutralization.

Subsequently, another study examined the effects of glucagon immunoneutralization, 2–4 hours after injection, in non-diabetic, moderately type 2 diabetic and severely type 1 diabetic alloxan (ALX)-induced rabbits. In non-diabetic rabbits, administration of monoclonal antibodies to glucagon decreased plasma glucose from ~7 to ~5 mmol/liter within 2 hours. In the ALX-induced Type 2 diabetic rabbits, immunoneutralization of glucagon essentially normalized fasting glucose levels, and in the ALX-induced Type 1 diabetic rabbits monoclonal antibody injection reduced fasting glucose levels from 21.2 to12.7 mmol/liter. Thus, interfering with glucagon action clearly was an effective method to reduce glucose levels in a rabbit model of Type 1 Diabetes Mellitus and to nearly normalize plasma glucose levels in a rabbit model of Type 2 Diabetes Mellitus.

Another recent study reported the effects of immunoneutralization of endogenous glucagon in obese hyperglycemic *ob/ob* mice. In this study, a single dose of a neutralizing monoclonal antibody decreased fasting glucose ≥ 40%, the glucose area under the curve (AUC) after an oral glucose load by ~50%, and fasting hepatic glucose production by ≥ 60%. Five-day treatment decreased mean blood glucose levels by 32%, and treatment with the monoclonal antibody for two weeks decreased HbA_{1C} by 1.0% relative to a control monoclonal antibody. However, fasting glucose levels were not reported (Heidi Sørensen et al. 2006).

Collectively, the findings from immunoneutralization studies indicate that glucagon contributes to hyperglycemia in diabetes and suggests that basal glucagon has a role in the maintenance of post-prandial plasma glucose levels. A relative excess of glucagon is a hallmark of diabetes mellitus, and several animal studies have demonstrated that interfering with glucagon action decreases fasting and post-prandial glucose levels in models of Type 2

Diabetes Mellitus. Hence, decreasing glucagon levels or action appears to be a promising therapeutic target for the management of Type 2 Diabetes Mellitus.

Therapeutic Approaches Targeting Glucagon Pathways for the Management of Diabetes

Peptides

The identification and glucose lowering effect of a peptide antagonist of the glucagon receptor was first reported more than twenty years ago (Johnson et al., 1984). In this study, the peptide antagonist (1-Nα-trinitrophenylhistidine, 12-homoarginine) glucagon, was administered to STZ-induced diabetic rats as a bolus of 1 mg/kg followed by continuous intravenous infusion at 33 μg/kg·min for 60 min. The peptide infusion was found to reduce blood glucose levels by ~67% compared to baseline at 5-minutes. Glycemic levels remained 30-55% below the baseline for more than 90 minutes. This finding provided the proof-of-concept that antagonizing glucagon action reduces plasma glucose levels. In the following years, extensive efforts were made to generate linear and cyclic peptide analogues with glucagon antagonistic activity that facilitates the inhibition glucagon stimulated cAMP production. One such compound, des-His1[Glu9]glucagon amide, was found to completely block the hyperglycemic effects of exogenous glucagon in normal rabbits (Unson et al., 1987) in addition to significantly improving hyperglycemia by a continuous inhibition of endogenous glucagon in streptozotocin (STZ)-induced diabetic rats (Unson et al., 1989).

Further, studies analyzing structure and function revealed that the N-terminal region of the glucagon receptor is primarily responsible for biological signaling and the C-terminal is important for ligand binding Hruby *et al.*, (1982). Considering that previous assays to detect adenylate cyclase were less specific, Van Tine *et al.*, (1996) used a phosphodiesterase inhibitor, Rolipram, to increase the sensitivity of this assay to allow identification of potent glucagon receptor antagonists. According to this study the previously discussed glucagon antagonist compounds were actually weak partial agonists, which demonstrated very low levels of cAMP accumulation that were sufficient to illicit an in vivo biological response. However, based on this revised adenylate cyclase assay, [des-His1, des-Phe6, Glu9]glucagon-NH_2, a newer glucagon antagonist, did not activate adenylate cyclase in the presence of Rolipram indicating that it was a pure antagonist of glucagon-induced adenylate cyclase activity. When administered to STZ-induced diabetic rats as a bolus of 0.75mg/kg followed by a continuous intravenous infusion at 25 μg/kg·min for 60 min, it reduced plasma glucose levels significantly by ~65% compared to baseline at 5 minutes and remained at this level throughout the 60-minute infusion.

The most recent evidence supporting the therapeutic potential of glucagon receptor peptide antagonists has been provided by a report assessing the peptidyl glucagon receptor antagonist, des-His-glucagon, in mice specifically engineered to express the human glucagon receptor instead of their endogenous mouse glucagon receptor (Dallas-Yang *et al.*, 2004). When administered intra-peritoneally at 2 mg/kg, this compound completely suppressed the increase in blood glucose induced by the subsequent intra-peritoneal administration of exogenous glucagon. Measurement of receptor occupancy revealed that ~80% of the hepatic glucagon receptors were occupied at 35 minutes after intra-peritoneal administration of the compound.

Non-Peptide, Small Molecules

Initial attempts to discover small molecule antagonists resulted in compounds with relatively reduced therapeutic potential. However, more recent reports have identified competitive and noncompetitive small molecule antagonists of the glucagon receptor. The compound NNC 25-2504 (3-cyano-4-hydroxybenzoic acid [1-(2,3,5,6-tetramethylbenzyl)-1H-indol-4-yl-methylene] hydrazide), a highly potent (IC50 = 2.3 nM), noncompetitive antagonist of the human glucagon receptor, inhibited glucagon-stimulated glucose production in isolated primary rat hepatocytes (Ki = 14 nM). In canine models, it was effective in suppressing the increase in blood glucose induced by the administration of exogenous glucagon (Ravi Kurukulasuriya & JT Link et al. 2005).

Another novel compound (N-[3-cyano-6-(1,1dimethylpropyl)-4,5,6,7-tetrahydro-1-benzothien-2-yl]-2-ethylbutanamide; compound 1) recently identified, was administered (15 or 50 mg/kg, ip) to mice engineered to express the human glucagon receptor in place of the endogenous mouse glucagon receptor (Qureshi SA et al. 2004). The administration of this compound was followed by an intra-peritoneal glucagon challenge administered one hour later. In mice treated with 50 mg/kg of this compound, the rapid glucagon-induced rise in blood glucose was significantly blunted. Interestingly, another study carried out using novel spiro-urea containing compounds, with the same experimental design and same animal model, also showed that these compounds when administered intra-peritoneally at 30mg/kg completely suppress the glucose excursion after the glucagon challenge (Shen et al. 2005).

One published report has shown that a small molecule non-peptide glucagon receptor antagonist can effectively suppress plasma glucose levels in humans (Peterson *et al.*, 2001). In this study, BAY 27-995 was administered at doses of 70mg and 200mg, and was found to effectively block the effects of exogenously administered glucagon in healthy males at a higher dose of 200mg. There was no demonstrable effect of this glucagon antagonist on fasting glucose levels, suggesting that in the absence of relative insulin deficiency, glucagon has no important role in determining post-prandial glucose levels or that a higher dose may be required to detect any effect. Given that BAY 27-9955 was well tolerated as a single dose and no serious clinical side effects were reported, the investigators concluded that this competitive glucagon antagonist might present a novel therapeutic approach to the treatment of Type 2 Diabetes. The therapeutic potential of such compounds in the management of Type 2 Diabetes currently remains undetermined.

Effect of Incretin-Based Therapies on Glucagon

GLP-1 is a 30-amino acid peptide produced by L cells in the lower intestine through alternative processing of the proglucagon gene product (Kieffer *et al.*, 1999). It is released in response to food intake and its most well-known physiological action is the 'incretin effect', which is mediated by enhancing glucose-induced insulin release (Kreyman *et al.*, 1987). Among the many mechanisms that contribute to its glucose lowering effect, GLP-1 also suppresses glucagon release in a glucose-dependent manner. Considering the therapeutic potential of this peptide, this action may be as important as its 'incretin effect' (Nuack *et al.*, 2002).

Gutniack *et al.*, (1992) assessed the effect of GLP-1 infusion on post-prandial glucagon levels in healthy subjects, patients with Type 1 and Type 2 Diabetes. They found that there was no glucagon response to meal ingestion and they suggested that the marked reduction in the post-prandial glucose excursion could be attributed to the insulinotropic action of GLP-1. Comparatively, in patients with Type 1 and Type 2 Diabetes there was a substantial increase in glucagon following meal intake. In patients with Type 2 Diabetes the infusion of GLP-1 was noted to decrease glucagon levels whereas in patients with Type 1 Diabetes it eliminated inappropriate elevation of glucagon after a meal, thereby reducing overall glucose excursions in both groups. Patients with diabetes in this study were receiving insulin via a closed-loop feedback system (Biostator). It is noteworthy that the amount of insulin required to maintain euglycemia was reduced by ~89% in patients with Type 2 Diabetes and ~50% in Type 1 Diabetes.

Further, Zander *et al.*, (2002) demonstrated that a continuous infusion of GLP-1 over 6 weeks reduced HbA1c by ~1.3% and resulted in improved β–cell function based on hyperglycemic clamp studies. This study also highlighted that in order to have a sustained therapeutic effect, GLP-1 must be continuously infused due to its short half-life (~1 min) as it is rapidly cleaved and inactivated by the enzyme dipeptidyl peptidase-4 (DPP-4), a post-proline cleaving serine protease. This observation resulted in extensive research leading to the development of several GLP-1 derivatives or mimetics in addition to pharmacological compounds that inhibit the action of DPP-4, which are currently approved for clinical use in the management of Type 2 Diabetes.

Exenatide, an incretin mimetic, is approved for use in patients with Type 2 Diabetes Mellitus as an adjunct to metformin, sulfonylureas, thiazolidinediones and/or basal insulin. It activates the GLP-1 receptor with a similar potency as native GLP-1; however, it is a poor substrate for DPP-4 as it has no alanine in the penultimate position (Barnett, 2005). Buse et al., (2004) showed that at week 30, treatment with Exenatide administered subcutaneously at doses of 5μg or 10μg twice daily significantly decreased HbA1c by 0.46 and 0.86% respectively. This significant reduction in HbA1c was achieved by reducing both fasting and post-prandial glucose concentrations. During phase 3 trials of glucagon, the effect of Exenatide on plasma glucagon levels was not documented, however, an earlier study demonstrated that Exenatide, like GLP-1, decreased fasting and post-prandial glucagon levels in healthy volunteers and in patients with Type 2 Diabetes (Nielson et al., 2004).

An important aspect of Exenatide's effect on glucagon secretion is that its actions are glucose-dependent. This has been demonstrated during stepped hypoglycemia clamps performed in healthy subjects where Exenatide significantly reduced plasma glucagon levels at glucose levels of 4.0Mm and 5.0mM, but it had no effect at glucose levels of 3.2mM. Surprisingly, it increased glucagon levels when glucose level reached 2.7mM (Degn *et al.*,

2004). Similarly, in a dose increasing design study in patients with Type2 Diabetes, 24-hour subcutaneous infusion of Exenatide at 0.2-0.8 μg/kg.d led to a significant decrease in the mean plasma glucagon levels throughout the day in addition to an increase in the insulin to glucagon ratio (Taylor *et al.*, 2005). Collectively, these clinical studies show that the therapeutic effect of GLP-1 mimetics in maintaining euglycemia is partly due to its effect on α–cell function.

Conclusion

Glucagon targeted therapies, being in practice since 2005 with the introduction of exenatide, has proved effective in treatment of type2 diabetes. Ongoing research to block glucagon at different stages has shown promising preliminary results.

References

Ahren B 2000 Autonomic regulation of islet hormone secretion – implications for health and disease. *Diabetologia* 43 393–410.

Barnett AH 2005 Exenatide. *Drugs Today (Barc)* 41:563–578

Baron AD, schaeffer L, Sharagg P, Kolterman OG. Role of hyperglucagonemia in maintenance of increased rates of hepatic glucose output in type2 diabetics. *Diabetes* 36:274-283, 1987.

Basu A, Shah P, Nielsen M, Basu R, Rizza RA 2004 Effects of type 2 diabetes on the regulation of hepatic glucose metabolism. *J Investig Med* 52:366–374

Basu R, Schwenk WF, Rizza RA 2004 Both fasting glucose production and disappearance are abnormal in people with "mild" and "severe" type 2 diabetes. *Am J Physiol Endocrinol Metab* 287:E55–E62

Beale E, Andreone T, Koch S, Granner M, and Granner D 1984 Insulin and glucagon regulate cytosolic phosphoenolpyruvate carboxykinase (GTP) mRNA in rat liver. *Diabetes* 33: 328 – 332

Bennett PH, Aronoff SL, Unger RH Evidence for an insulin-independent alpha-cell abnormality in human diabetes. *Metabolism* 25 (1976) (11 Suppl 1):1527–1529.

Beuers U and Jungermann K 1990 Relative contribution of glyco- genolysis and gluconeogenesis to basal, glucagon- and nerve stimulation-dependent glucose output in the perfused liver from fed and fasted rats. *Biochem Int* 21: 405–415

Bokvist K, Olsen HL, Hoy M, Gotfredsen CF, Holmes WF, Buschard K, Rorsman P & Gromada J 1999 Characterisation of sulphonylurea and ATP-regulated K+ channels in rat pancreatic A-cells. *Pflugers Archiv: European Journal of Physiology* 438 428–436.

Brelje TC, Scharp DW & Sorenson RL 1989 Three-dimensional imaging of intact isolated islets of Langerhans with confocal microscopy. *Diabetes* 38 808–814.

Brissova M, Fowler MJ, Nicholson WE, Chu A, Hirshberg B, Harlan DM & Powers AC 2005 Assessment of human pancreatic islet architecture and composition by laser scanning confocal microscopy. *Journal of Histochemistry and Cytochemistry* 53 1087–1097.

Burcelin R, Katz EB, and Charron MJ. 1996 Molecular and cellu- lar aspects of the glucagon receptor: role in diabetes and metabolism. *Diabetes Metab* 22: 373–396

Buse JB, Henry RR, Han J, Kim DD, Fineman MS, Baron AD 2004 Effects of exenatide (exendin-4) on glycemic control over 30 weeks in sulfonylurea-treated patients with type 2 diabetes. *Diabetes Care* 27:2628–2635

Cabrera O, Berman DM, Kenyon NS, Ricordi C, Berggren PO & Caicedo A 2006 The unique cytoarchitecture of human pancreatic islets has implications for islet cell function. *PNAS* 103 2334–2339.

Castadelli A, Baldi S, Pettiti M, Toschi E. influence of obesity and type2 diabetes on gluconeogenisis and glucose output in human: a quantitative study. *Diabetes* 49:1367-1373, 2000.

Castano JG, Nieto A, and Feliu JE. 1979 Inactivation of phospho- fructokinase by glucagon in rat hepatocytes. *J Biol Chem* 254: 5576–5579.

Chen M, Gavrilova O, Zhao WQ, Nguyen A, Lorenzo J, Shen L, Nackers L, Pack S, Jou W, Weinstein LS 2005 Increased glucose tolerance and reduced adiposity in the absence of fasting hypoglycemia in mice with liver-specific Gs α-deficiency. *J Clin Invest* 115:3217–3227

Christ B, Nath A, Bastian H, and Jungermann K 1988 Regula- tion of the expression of the phosphoenolpyruvate carboxykinase gene in cultured rat hepatocytes by glucagon and insulin. *Eur J Biochem* 178: 373 – 379

Christophe J. 1996 Glucagon and its receptor in various tissues. *Ann NY Acad Sci* 805: 31–43

Clarke WL, Gonder-Frederick LA, Richards FE, Cryer PE Multifactorial origin of hypoglycemic symptom unawareness in IDDM. Association with defective glucose counterregulation and better glycemic control. *Diabetes* (1991) 40: 680 – 685.

Consoli A, Nurjhan N, Capani F, Gerich J 1989 Predominant role of gluconeogenesis in increased hepatic glucose production in NIDDM. *Diabetes* 38:550–557

Dale S. Edgerton and Alan D. Cherrington 2011. Glucagon as a Critical Factor in the Pathology of Diabetes. *DIABETES* 60: 377-380.

Dallas-Yang Q, Shen X, Strowski M, Brady E, Saperstein R, Gibson RE, Szalkowski D, Qureshi SA, Candelore MR, Fenyk-Melody JE, Parmee ER, Zhang BB, Jiang G 2004 Hepatic glucagon receptor binding and glucose-lowering in vivo by peptidyl and non-peptidyl glucagon receptor antagonists. *Eur J Pharmacol* 501:225–234

DeFronzo RA, Ferrannini E, Simonson DC 1989 Fasting hyperglycemia in non-insulin-dependent diabetes mellitus: contributions of excessive hepatic glucose production and impaired tissue glucose uptake. *Metabolism* 38:387–395

Degn KB, Brock B, Juhl CB, Djurhuus CB, Grubert J, Kim D, Han J, Taylor K, Fineman M, Schmitz O 2004 Effect of intravenous infusion of exenatide (synthetic exendin-4) on glucose-dependent insulin secretion and counterregulation during hypoglycemia. *Diabetes* 53:2397–2403

Dey A, Lipkind GM, Rouille Y, Norrbom C, Stein J, Zhang C, Carroll R & Steiner DF 2005 Significance of prohormone convertase 2, PC2, mediated initial cleavage at the proglucagon interdomain site, Lys70-Arg71, to generate glucagon. *Endocrinology* 146 713–727.

Dinneen S, Alzaid A, Turk D, Rizza R 1995 Failure of glucagon suppression contributes to postprandial hyperglycaemia in IDDM. *Diabetologia* 38:337–343

Dobbins RL, Davis SN, Neal D, Caumo A, Cobelli C, and Cherrington AD 1998 Rates of glucagon activation and deactivation of hepatic glucose production in conscious dogs. *Metabolism* 47: 135–142.

Dumonteil E, Magnan C, Ritz-Laser B, Ktorza A, Meda P & Philippe J 2000 Glucose regulates proinsulin and prosomatostatin but not proglucagon messenger ribonucleic acid levels in rat pancreatic islets. *Endocrinology* 141 174–180

Dumonteil E, Magnan C, Ritz-Laser B, Meda P, Dussoix P, Gilbert M, Ktorza A & Philippe J 1998 Insulin, but not glucose lowering corrects the hyperglucagonemia and increased proglucagon messenger ribonucleic acid levels observed in insulinopenic diabetes. *Endocrinology* 139 4540–4546.

Felig P, Wahren J, Hendler R 1976 Influence of physiologic hyperglucagonemia on basal and insulin-inhibited splanchnic glucose output in normal man. *J Clin Invest* 58:761–765

Franklin I, Gromada J, Gjinovci A, Theander S, Wollheim CB. 2005 Beta-cell secretory products activate alpha-cell ATP-dependent potas- sium channels to inhibit glucagon release. *Diabetes* 54: 1808–1815

Freychet L, Rizkalla SW, Desplanque N, Basdevant A, Zirinis P, Tchobroutsky G, and Slama G. 1988 Effect of intranasal glucagon on blood glucose levels in healthy subjects and hypo- glycaemic patients with insulin-dependent diabetes. *Lancet* 1: 1364–1366

Furuta M, Zhou A, Webb G, Carroll R, Ravazzola M, Orci L & Steiner DF 2001 Severe defect in proglucagon processing in islet A-cells of prohormone convertase 2 null mice. *Journal of Biological Chemistry* 276 27197–27202.

Gelling RW, Du XQ, Dichmann DS, Romer J, Huang H, Cui L, Obici S, Tang B, Holst JJ, Fledelius C, Johansen PB, Rossetti L, Jelicks LA, Serup P, Nishimura E, Charron MJ 2003 Lower blood glucose, hyperglucagonemia, and pancreatic α-cell hyperplasia in glucagon receptor knockout mice. *Proc Natl Acad Sci USA* 100:1438–1443

Gerich J, Mara Lorenzi, John H Karam; Victor Schneider, Pete Forsham. Abnormal pancreatic glucagon secretion and postprandial hyperglycemia in Diabetes mellitus. JAMA (1975)-vol 234,2: 159- 165.

Gerich JE 1997 Metabolic abnormalities in impaired glucose tolerance. *Metabolism* 46:40–43

Gerich JE, Charles MA, Grodsky GM 1976 Regulation of pancreatic insulin and glucagon secretion. *Ann Rev Physiol* 38:353–388

Gerich JE, Lorenzi M, Bier DM, Schneider V, Tsalikian E, Karam JH, Forsham PH 1975 Prevention of human diabetic ketoacidosis by somatostatin. Evidence for an essential role of glucagon. *N Engl J Med* 292:985–989

Gerich JE, Lorenzi M, Bier DM, Tsalikian E, Schneider V, Karam JH, Forsham PH 1976 Effects of physiologic levels of glucagon and growth hormone on human carbohydrate and lipid metabolism. Studies involving administration of exogenous hormone during suppression of endogenous hormone secretion with somatostatin. *J Clin Invest* 57:875–884

Gerich JE, Schneider VS, Lorenzi M, Tsalikian E, Karam JH, Bier DM, Forsham PH 1976 Role of glucagon in human diabetic ketoacidosis: studies using somatostatin. *Clin Endocrinol* (Oxf) 5(Suppl):299S–305S

Gopel SO, Kanno T, Barg S & Rorsman P 2000 Patch-clamp characterisation of somatostatin-secreting delta-cells in intact mouse pancreatic islets. *Journal of Physiology* 528 497–507.

Greenbaum CJ, Prigeon RL, D'Alessio DA 2002 Impaired □-cell function, incretin effect, and glucagon suppression in patients with type 1 diabetes who have normal fasting glucose. *Diabetes* 51:951– 957

Gromada J, Bokvist K, Ding WG, Barg S, Buschard K, Renstrom E & Rorsman P 1997 Adrenaline stimulates glucagon secretion in pancreatic A-cells by increasing the Ca2+ current and the number of granules close to the L-type Ca2+ channels. *Journal of General Physiology* 110 217–228.

Gromada J, Franklin I & Wollheim CB 2007 Alpha-cells of the endocrine pancreas: 35 years of research but the enigma remains. *Endocrine Reviews* 28 84–116.

Gromada J, Ma X, Hoy M, Bokvist K, Salehi A, Berggren PO & Rorsman PATP-sensitive K+ channel-dependent regulation of glucagon release and electrical activity by glucose in wild-type and SUR1−/− mouse alpha-cells*Diabetes* 53 Suppl 3 2004 S181–S189.

Gutniak M, Ørskov C, Holst JJ, Ahrén B, Efendic S 1992 Antidiabetogenic effect of glucagon-like peptide-1 (7-36) amide in normal subjects and patients with diabetes mellitus. *N Engl J Med* 326:1316–1322

Hansen LH, Abrahamsen N, Nishimura E. 1995 Glucagon receptor mRNA distribution in rat tissues. *Peptides* 16: 1163–1166

Heidi Sørensen, Christian L. Brand, Susanne Neschen, Jens Juul Holst, Keld Fosgerau, Erica Nishimura, and Gerald I. Shulman. Immunoneutralization of Endogenous Glucagon Reduces Hepatic Glucose Output and Improves Long-Term Glycemic Control in Diabetic ob/ob Mice. *Diabetes* 55:2843–2848, 2006

Herzig S, Long F, Jhala US, Hedrick S, Quinn R, Bauer A, Rudolph D, Schutz G, Yoon C, Puigserver P, Spiegelman B, and Montminy M. 2001 CREB regulates hepatic gluconeogenesis through the coactivator PGC-1. *Nature* 413: 179 – 183

Hruby VJ 1982 Structure-conformation-activity studies of glucagon and semi-synthetic glucagon analogs. *Mol Cell Biochem* 44:49–64

Hvidberg A, Djurup R, and Hilsted J. 1994 Glucose recovery after intranasal glucagon during hypoglycaemia in man. *Eur J Clin Pharmacol* 46: 15–17

Ikeda T, Hoshino T, Honda M, Takeuchi T, Mokuda O, Tominaga M, and Mashiba H. 1989 Effect of glucagon on glucose output from bivascularly perfused rat liver. *Exp Clin Endocrinol* 94: 383–386

Jelinek LJ, Lok S, Rosenberg GB, Smith RA, Grant FJ, Biggs S, Bensch PA, Kuijper JL, Sheppard PO, and Sprecher CA. 1993 Expression cloning and signaling properties of the rat glucagon receptor. *Science* 259: 1614–1616

Jiang G, Zhang BB. 2003 Glucagon and regulation of glucose metabolism. *Am J Physiol Endocrinol Metab* 284: E671–E678

Johnson LN, Barford D, Owen DJ, Noble ME, and Garman EF. 1997 From phosphorylase to phosphorylase kinase. *Adv Second Messenger Phosphoprotein Res* 31: 11–28

Kieffer TJ, Habener JF 1999 The glucagon-like peptides. *Endocr Rev* 20:876–913

Komjati M, Bratusch-Marrain P, and Waldhausl W. 1986 Superior efficacy of pulsatile versus continuous hormone exposure on hepatic glucose production in vitro. *Endocrinology* 118: 312–319

Krebs EG. 1981 Phosphorylation and dephosphorylation of glycogen phosphorylase: a prototype for reversible covalent enzyme modification. *Curr Top Cell Regul* 18: 401–419

Kreymann B, Williams G, Ghatei MA, Bloom SR 1987 Glucagon-like peptide-1 7-36: a physiological incretin in man. *Lancet* 2:1300–1304

Lee Y, Wang MY, Du XQ, Charron MJ, Unger RH 2011. Glucagon receptor knockout prevents insulin-deficient type1 diabetes in mice. Diabetes 60:391–397

Liang Y, Osborne MC, Monia BP, Bhanot S, Gaarde WA, Reed C, She P, Jetton TL, Demarest KT 2004 Reduction in glucagon receptor expression by an antisense oligonucleotide ameliorates diabetic syndrome in *db/db* mice. *Diabetes* 53:410–417

Lins PE, Wajngot A, Adamson U, Vranic M, Efendic S 1983 Minimal increases in glucagon levels enhance glucose production in man with partial hypoinsulinemia. *Diabetes* 32:633–636

MacDonald PE, De Marinis YZ, Ramracheya R, Salehi A, Ma X, Johnson PR, Cox R, Eliasson L & Rorsman P 2007 A K ATP channel-dependent pathway within alpha cells regulates glucagon release from both rodent and human islets of Langerhans. *PLoS Biology* 5 e143.

Magnusson I, Rothman DL, Katz LD, Shulman RG, Shulman GI 1992 Increased rate of gluconeogenesis in type II diabetes mellitus. A 13C nuclear magnetic resonance study. *J Clin Invest* 90:1323–1327

Marks JS and Botelho LH. 1986 Synergistic inhibition of glucagon- induced effects on hepatic glucose metabolism in the presence of insulin and a cAMP antagonist. *J Biol Chem* 261: 15895–15899

Mitrakou A, Kelley D, Mokan M, Veneman T, Pangburn T, Reilly J, Gerich J 1992 Role of reduced suppression of glucose production and diminished early insulin release in impaired glucose tolerance. *N Engl J Med* 326:22–29.

Mitrakou A, Kelley D, Veneman T, Jenssen T, Pangburn T, Rei, lly J, Gerich J 1990 Contribution of abnormal muscle and liver glucose metabolism to postprandial hyperglycemia in NIDDM. *Diabetes* 39:1381–1390.

Mojsov S, Heinrich G, Wilson IB, Ravazzola M, Orci L & Habener JF 1986 Preproglucagon gene expression in pancreas and intestine diversifies at the level of post-translational processing. *Journal of Biological Chemistry* 261 11880–11889.

Muller WA, Faloona GR, Unger RH The effect of experimental insulin deficiency on glucagon secretion. *J Clin Invest* (1971) 50:1992–1999.

Myers SR, Diamond MP, Adkins-Marshall BA, Williams PE, Stinsen R, and Cherrington AD. 1991 Effects of small changes in glucagon on glucose production during a euglycemic, hyper- insulinemic clamp. *Metabolism* 40: 66–71

Nadal A, Quesada I & Soria B 1999 Homologous and heterologous asynchronicity between identified alpha-, beta- and delta-cells within intact islets of Langerhans in the mouse. *Journal of Physiology* 517 85–93.

Nielsen LL, Young AA, Parkes DG 2004 Pharmacology of exenatide (synthetic exendin-4): a potential therapeutic for improved glycemic control of type 2 diabetes. *Regul Pept* 117:77–88

Okar DA and Lange AJ. 1999 Fructose-2,6-bisphosphate and con- trol of carbohydrate metabolism in eukaryotes. *Biofactors* 10: 1–14

Olsen HL, Theander S, Bokvist K, Buschard K, Wollheim CB & Gromada J 2005 Glucose stimulates glucagon release in single rat alpha-cells by mechanisms that mirror the stimulus-secretion coupling in beta-cells. *Endocrinology* 146 4861–4870.

Opara EC, Atwater I, and Go VL. 1988 Characterization and con- trol of pulsatile secretion of insulin and glucagon. *Pancreas* 3: 484–487

Paolisso G, Scheen AJ, Albert A, and Lefebvre PJ. 1989 Effects of pulsatile delivery of insulin and glucagon in humans. *Am J Physiol Endocrinol Metab* 257: E686–E696

Parker JC, Andrews KM, Allen MR, Stock JL, McNeish JD 2002 Glycemic control in mice with targeted disruption of the glucagon receptor gene. *Biochem Biophys Res Commun* 290:839–843

Petersen KF, Sullivan JT 2001 Effects of a novel glucagon receptor antagonist (Bay 27-9955) on glucagon-stimulated glucose production in humans. *Diabetologia* 44:2018–2024

Philippe J, Morel C & Cordier-Bussat M 1995 Islet-specific proteins interact with the insulin-response element of the glucagon gene. *Journal of Biological Chemistry* 270 3039–3045.

Quesada I, Todorova MG & Soria B 2006a Different metabolic responses in α- β-, and δ-cells of the islet of Langerhans monitored by redox confocal microscopy. *Biophysical Journal* 90 2641–2650

Qureshi SA, Rios CM, Xie D, Yang X, Tota LM, Ding VD, Li Z, Bansal A, Miller C, Cohen SM, Jiang G, Brady E, Saperstein R, Duffy JL, Tata JR, Chapman KT, Moller DE, Zhang BB 2004 A novel glucagon receptor antagonist inhibits glucagon-mediated biological effects. *Diabetes* 53:3267–3273

Ramachandran C, Angelos KL, and Walsh DA. 1983 Hormonal regulation of the phosphorylation of glycogen synthase in per- fused rat heart. Effects of insulin, catecholamines, and glucagon. *J Biol Chem* 258: 13377 – 13383.

Ravi Kurukulasuriya & JT Link. Progress towards glucagon receptor antagonist therapy for Type 2 diabetes. *Expert Opin. Ther. Patents* (2005) 15(12):1739-1749

Roach PJ. 1990 Control of glycogen synthase by hierarchal protein phosphorylation. *FASEB J* 4: 2961 – 2968

Rouille Y, Westermark G, Martin SK, Steiner DF 1994 Proglucagon is processed to glucagon by prohormone convertase PC2 in αTC1–6 cells. *Proc Natl Acad Sci USA* 91:3242–3246

Shah P, Vella A, Basu A, Basu R, Schwenk WF, Rizza RA 2000 Lack of suppression of glucagon contributes to postprandial hyperglycemia in subjects with type 2 diabetes mellitus. *J Clin Endocrinol Metab* 85:4053–4059

Shen DM, Zhang F, Brady EJ, Candelore MR, las-Yang Q, Ding VD, Dragovic J, Feeney WP, Jiang G, McCann PE, Mock S, Qureshi SA, Saperstein R, Shen X, Tamvakopoulos C, Tong X, Tota LM, Wright MJ, Yang X, Zheng S, Chapman KT, Zhang BB, Tata JR, Parmee ER 2005 Discovery of novel, potent, and orally active spiro-urea human glucagon receptor antagonists. *Bioorg Med Chem* 15:4564 – 4569

Sherwin RS, Fisher M, Hendler R, Felig P 1976 Hyperglucagonemia and blood glucose regulation in normal, obese and diabetic subjects. *N Engl J Med* 294:455–461

Sloop KW, Cao JX, Siesky AM, Zhang HY, Bodenmiller DM, Cox AL, Jacobs SJ, Moyers JS, Owens RA, Showalter AD, Brenner MB, Raap A, Gromada J, Berridge BR, Monteith DK, Porksen N, McKay RA, Monia BP, Bhanot S, Watts LM, Michael MD 2004 Hepatic and glucagon-like peptide-1-mediated reversal of diabetes by glucagon receptor antisense oligonucleotide inhibitors. *J Clin Invest* 113:1571–1581

Stevenson RW, Steiner KE, Davis MA, Hendrick GK, Williams PE, Lacy WW, Brown L, Donahue P, Lacy DB, Cherrington AD 1987 Similar dose responsiveness of hepatic glycogenolysis and gluconeogenesis to glucagon in vivo. *Diabetes* 36:382–389

Striffler JS, Garfield SA, Cardell EL, and Cardell RR. 1984 Effects of glucagon on hepatic microsomal glucose-6-phosphatase in vivo. *Diabetes Metab* 10: 91 – 97

Taylor K, Kim D, Nielsen LL, Aisporna M, Baron AD, Fineman MS 2005 Day-long subcutaneous infusion of exenatide lowers glycemia in patients with type 2 diabetes. *Horm Metab Res* 37:627–632

Unson CG, Andreu D, Gurzenda EM, Merrifield RB 1987 Synthetic peptide antagonists of glucagon. *Proc Natl Acad Sci USA* 84:4083–4087

Unson CG, Gurzenda EM, Merrifield RB 1989 Biological activities of des-His1[Glu9]glucagon amide, a glucagon antagonist. *Peptides* 10:1171–1177

Van Tine BA, Azizeh BY, Trivedi D, Phelps JR, Houslay MD, Johnson DG, Hruby VJ 1996 Low level cyclic adenosine 3′,5′-monophosphate accumulation analysis of [des-His1, des- Phe6, Glu9] glucagon-NH2 identifies glucagon antagonists from weak partial agonists/antagonists. *Endocrinology* 137:3316–3322

Veneziale CM, Deering NG, and Thompson HJ. 1976 Gluconeogenesis in isolated rat hepatic parenchymal cells. IX. Differential effects of glucagon and epinephrine on phosphofructokinase and pyruvate kinase. *Mayo Clin Proc* 51: 624 – 631

Vieira E, Salehi A & Gylfe E 2007 Glucose inhibits glucagon secretion by a direct effect on mouse pancreatic alpha cells. *Diabetologia* 50 370–379.

Vozzi C, Ullrich S, Charollais A, Philippe J, Orci L & Meda P 1995 Adequate connexin-mediated coupling is required for proper insulin production. *Journal of Cell Biology* 131 1561–1572.

Weigle DS and Goodner CJ. 1986 Evidence that the physiological pulse frequency of glucagon secretion optimizes glucose production by perifused rat hepatocytes. *Endocrinology* 118: 1606–1613

Weigle DS, Koerker DJ, and Goodner CJ. 1984 Pulsatile glucagon delivery enhances glucose production by perifused rat hepatocytes. *Am J Physiol Endocrinol Metab* 247: E564 – E568

Yoon JC, Puigserver P, Chen G, Donovan J, Wu Z, Rhee J, Adelmant G, Stafford J, Kahn CR, Granner DK, Newgard CB, and Spiegelman BM. 2001 Control of hepatic gluconeogenesis through the transcriptional coactivator PGC-1. *Nature* 413: 131– 138

Young AA, Cooper GJ, Carlo P, Rink TJ, and Wang MW. 1993 Response to intravenous injections of amylin and glucagon in fasted, fed, and hypoglycemic rats. *Am J Physiol Endocrinol Metab* 264: E943 – E950

Zander M, Madsbad S, Madsen JL, Holst JJ 2002 Effect of 6-week course of glucagon-like peptide 1 on glycaemic control, insulin sensitivity, and β-cell function in type 2 diabetes: a parallel-group study. *Lancet* 359:824–830

Zhou H, Zhang T, Harmon JS, Bryan J & Robertson RP 2007a Zinc, not insulin, regulates the rat alpha-cell response to hypoglycemia *in vivo. Diabetes* 56 1107–1112.

In: Glucagon
Editor: Ahmed Al-Jebawi

ISBN: 978-1-62618-147-2

Chapter 7

Alpha-Cell-Mediated Beta-Like Cell Regeneration as a Putative Therapeutic Pathway for Type 1 Diabetes

Anja Pfeifer[1,2], Keith Al-Hasani[1,2], Monica Courtney[1,2], Nouha Ben-Othman[1,2], Elisabet Gjernes[1,2], Andhira Vieira[1,2] and Patrick Collombat*[1,2]

[1] INSERM U636, Diabetes Genetics Team, Nice, France

[2] Université de Nice-Sophia Antipolis, Nice, France

Abstract

The human body requires the maintenance of stable and balanced blood glucose levels. This state of homeostasis is strictly regulated by the antagonistic actions of pancreatic peptide hormones, including insulin and glucagon. The mature pancreas consists of two compartments with distinct functions: the exocrine tissue, where digestive enzymes are produced and transported into the duodenum via a widely branched ductal system, and the endocrine compartment that consists of highly vascularized cell-agglomerates, called islets of Langerhans. These functional units include insulin-producing beta-cells and glucagon-

* To whom correspondence should be addressed: collombat@unice.fr.

secreting alpha-cells. Specifically, upon increased blood sugar levels, insulin is released and induces fat-, muscle-, and red blood cells, to store and/or metabolize glucose. Conversely, glucagon, which is released by alpha-cells in response to low levels of blood glucose, predominantly promotes liver cells to release/synthesize glucose. Should the strictly regulated interplay between insulin and glucagon be impaired, the required narrow range of blood glucose levels can no longer be preserved, and severe diseases, such as diabetes, may arise. Diabetes mellitus, which is the result of a cell-mediated autoimmune destruction of beta-cells, has become one of the most widespread metabolic diseases, affecting almost 6% of the world's population. The absolute or relative deficiency of insulin results in the development of Type 1 Diabetes Mellitus (T1DM) or Type 2 Diabetes Mellitus (T2DM), respectively.

Although insulin-based therapy can provide a measure of control of the glycemia in T1DM patients, exogenous insulin can still not ensure physiologically stable blood glucose levels. Differences in diet, age and/or exercise can indeed cause significant variations in the glycemia. Such fluctuations represent a major burden, since they can, over time, lead to a wide range of complications, including micro- and macro-vascular damages, blindness, amputation and/or death. Alternative therapies, such as islet transplantation or *in vitro* generation of beta-cells, present a number of disadvantages, including life-long immunosuppression, the scarcity of cadaveric donors, or technical inefficiencies.

Importantly, using the mouse as a model, a new alternative was recently described *in vivo*. Indeed, it was demonstrated that glucagon-producing alpha-cells could be continuously regenerated and subsequently converted into functional insulin-expressing beta-like cells, such cells being capable of reversing the consequences of chemically-induced hyperglycemia. This chapter will therefore focus on the potential of alpha-to-beta-like cell transdifferentiation as a therapy for T1DM.

1. Introduction

Diabetes Mellitus is one of the most common metabolic diseases with a prevalence of approximately 300 million people worldwide. It encompasses two related, but distinct, subtypes [1]. Type 1 Diabetes Mellitus (T1DM) is caused by a cell-specific autoimmune loss of insulin-expressing beta-cells in the pancreatic islets of Langerhans [2]. Type 2 Diabetes Mellitus (T2DM) is due to a progressive failure in insulin production/function. Although insulin-based therapy can provide a measure of control of the glycemia in patients with T1DM, significant variations may still occur in time due to certain environmental factors, including diet, age, and/or exercise. Such fluctuations represent a major burden since they can, over time, lead to a wide range of micro- and macro-vascular damages including blindness, amputation, and/or premature death [3]. An alternative to the classical insulin supplementation is cadaveric islet transplantation. However, due to the inherent life-long immunosuppressive treatment and scarcity of donors, this approach is rarely used [4]. It is important to notice that despite current therapies, patients with T1DM tend to have a shorter life expectancy compared to their healthy counterparts. Interestingly, using the mouse as a model, a new approach in the context of T1DM was recently described. Through this model, it was demonstrated that glucagon-producing alpha-cells can be continuously regenerated and converted into functional insulin-expressing beta-like cells [5]. These cells are capable of reversing the consequences of chemically-induced hyperglycemia. This chapter will therefore focus on these findings and assess the potential of such alpha-to-beta-like cell conversion and the new research avenues it may open towards a putative treatment of T1DM.

2. Pancreas Anatomy and Function

The pancreas can be subdivided into the head region, located across the back of the abdomen, and the tail portion, found near the spleen. The pancreas is composed of two distinct compartments: the exocrine compartment, encompassing acinar and duct cells, and the endocrine tissue. Acinar cells secrete a number of digestive enzymes, such as amylase and trypsin, into a network of ducts that reach the main pancreatic duct along the organ. Those digestive enzymes travel through the main pancreatic duct into the bile duct in an inactive form (zymogen). Subsequently, they are activated in the duodenum where they contribute to the processing of carbohydrates, proteins, fats, and acids. The endocrine tissue secretes several hormones that are directly released into the bloodstream; it is organized into clusters of cells termed islets of Langerhans [6]. These highly vascularized cell clusters, first observed by Paul Langerhans in the late 19th century [7], are scattered throughout the pancreatic tissue and make up less than 2% of the organ's total mass. They include five distinct cell subtypes, alpha-, beta-, pancreatic polypeptide (PP), delta- and epsilon-cells, that are all associated with the secretion of specific endocrine hormones. Insulin and glucagon are secreted by beta- and alpha-cells, respectively. They display antagonistic functions in the regulation and maintenance of physiologically stable blood glucose levels. Insulin is released upon increased blood glucose levels and induces adipose, muscle, and hepatic cells to store and/or metabolize glucose. Conversely, glucagon is released in response to hypoglycemia and acts by promoting hepatic cells to release glucose through increasing both glycogenolysis and gluconeogenesis. Another peptide hormone, somatostatin, produced by delta-cells, prevents the release of both insulin and glucagon [8]. Pancreatic polypeptide (PP) is secreted by PP-cells and is involved in the regulation of pancreatic secretion activities [9]. Murine islets of Langerhans are organized in a very specific fashion (Fig. 2A). Insulin-expressing beta-cells, which make up approximately 75% of the islet cell mass, are located within the core of the islets, whereas glucagon-expressing alpha-cells (about 20% of islet cells), are predominantly detected on the mantle of the islets.

3. Pancreatic Development

In mice, the first morphological signs of pancreatic development can be detected at approximately embryonic day 9 (E9) with the emergence of two epithelial buds from both the dorsal and ventral sides of the gut epithelium. Prior to and during budding, the organ primordium expresses a number of factors that promote the specification of the endodermal cells towards the pancreatic lineage, including the homeodomain-containing protein Pdx1. *Pdx1* is broadly expressed until E13 when it becomes restricted to beta-cells and some delta-cells, although low expression of *Pdx1* is maintained in a subset of duct or acinar cells [10].

In adult pancreata, *Pdx1* expression is restricted to the ductal epithelium and beta-cells [11]. Lineage tracing studies have shown that all pancreatic cells are derived from $Pdx1^{+}$ progenitors [12, 13]. The finding that pancreatic development is arrested at the budding stage in *Pdx1* mutants was quite interesting because it further demonstrated the crucial role exerted by this factor for proper pancreas development [14, 15].

A key regulator of endocrine development is the proendocrine basic helix-loop-helix (bHLH)- containing transcription factor Neurogenin3 (*Ngn3*), which is expressed exclusively in endocrine precursor cells, and is subsequently down-regulated during pancreas morphogenesis [13, 16, 17]. In *Ngn3*-KO mice, the pancreas is properly formed, but lacks endocrine cells, supporting the notion that Ngn3 plays a pivotal role in the specification towards the endocrine cell lineage.

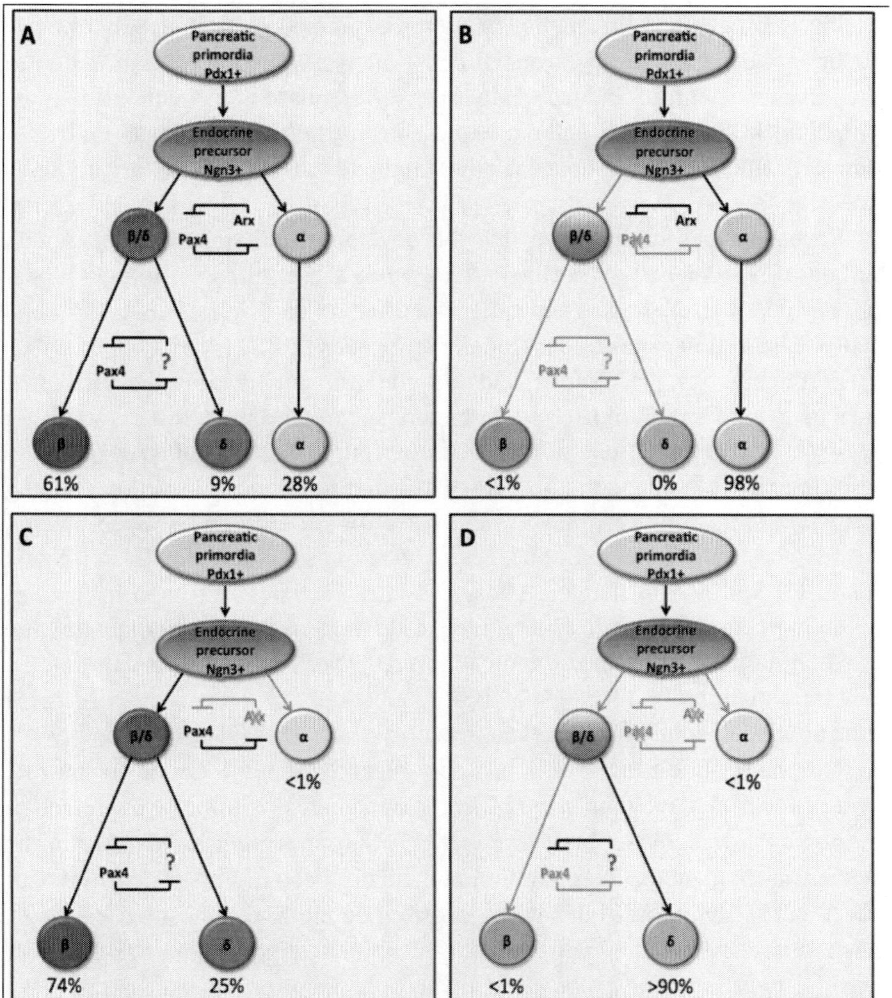

Figure 1. Schematic model illustrating the specification pathways of endocrine cell subtypes in wild-type, *Arx*-, *Pax4*-, and *Arx*-/*Pax4*-deficient animals. Upon the expression of a number of factors, including Pdx1 and Ngn3, pancreatic precursor cells are specified towards the pancreatic and endocrine cell fates, respectively. In wild-type animals, $Ngn3^+$ cells adopt either a beta/delta or an alpha cell-fate, depending on the mutual inhibitory actions of Arx and Pax4 (A). In *Pax4*-deficient mice, a nearly exclusive generation of alpha-cells is detected (B), whereas the deletion of the *Arx* gene leads to the generation of beta- and delta-cells, concomitantly with an almost complete loss of alpha-cells (C). The deficiency in both transcription factors results in the generation of delta-cells (D), indicating that a continuous predominance of *Pax4* expression is required in a second round of specification, in order to direct beta/delta precursor cells towards a mature beta-cell fate.

$Pdx1^+$ progenitors in which *Ngn3* is not expressed acquire an exocrine cell fate and give rise to acinar or ductal cells. Interestingly, the misexpression of *Ngn3* is sufficient to induce endocrine differentiation throughout the gut epithelium [18], while its ectopic expression in the early pancreas under the control of the *Pdx1* promoter results in the diversion of the organ to an endocrine fate [16, 19].

Recent studies have shown that the developmental stage correlates with the fate of Ngn3-labelled precursor cells: while these initially adopt an alpha-cell destiny, these are subsequently specified towards the beta-, PP-, and finally delta-cell lineages as development proceeds [20].

Following Ngn3 activation and the commitment towards the endocrine lineage, a number of transcription factors are required in order to further specify the differentiation into one of the four islet cell subtypes. Among these, the paired homeodomain factor Pax4 and the homeodomain-containing factor Arx were found to play crucial roles for the differential specification towards the four endocrine lineages. The expression of *Pax4* is initiated around E9.5 in both pancreatic buds, and later becomes restricted to insulin-expressing beta-cells. In *Pax4* KO mice, a dramatic reduction in the beta- and delta-cell numbers is detected, indicating that *Pax4* expression is required for the differentiation of both cell subtypes. Interestingly, a compensatory increase in the alpha-cell content is observed in such animals [21] (Fig. 1B).

The transcription factor Arx has only recently been described for its role during endocrine development [22]. In the pancreas, it is initially expressed at the onset of pancreatic bud evagination. Subsequently, its expression is detected in the pancreatic epithelium and finally becomes restricted to alpha- and PP-cells. Mutations in this gene, localized on the X chromosome, cause X-linked retardation and epilepsy and/or deregulated glycemia. *Arx*-deficient mice die two days *postpartum* and display dehydration, weakness and very low blood glucose levels. Interestingly, in *Arx*-deficient animals, endocrine precursors exclusively adopt a beta- or delta-cell fate, resulting in the loss of alpha-cells (Fig. 1C). These findings imply that *Arx* acts as an antagonist of *Pax4* (and *vice versa*) in the processes underlying the specification towards the different endocrine cell lineages. In fact, the two factors were shown to be initially co-expressed and to interact with each other's promoters through an inhibitory cross-regulatory circuit that controls the transcriptional state of these two genes [23]. Accordingly, *Arx* or *Pax4* transcripts are found up-regulated in mice deficient for the antagonistic gene. Thus, during the course of pancreas morphogenesis, if *Pax4* predominates, progenitor cells adopt a beta- or delta-cell fate, whereas *Arx* domination leads to specification towards the alpha-cell lineage. Importantly, the analysis of mice deficient for both *Arx* and *Pax4* brought further insight into the transcriptional pathways that regulate endocrine progenitor cell specification [23]: an almost complete loss of both beta- and alpha-cells and a nearly exclusive generation of delta-cells was detected (Fig. 1D). These findings suggest that endocrine precursor cells undergo at least two rounds of specification: first, they are either directed towards the beta/delta-cell lineage (upon *Pax4* predomination) or the alpha-cell fate (if *Arx* predominates). If *Pax4* expression continuously prevails over a hitherto unidentified factor in a second round of specification, the beta-cell fate is finally chosen at the expense of the delta-cell lineage.

4. *In Vivo* Conversion of Glucagon-Expressing Alpha-Cells into Insulin-Expressing Beta-Like Cells

Type 1 diabetes is caused by an autoimmune-mediated loss of insulin-expressing beta-cells resulting in chronic hyperglycemia. Insulin-based therapy can provide a measure of control of blood glucose levels, but it fails to mimic normal insulin physiology and therefore, does not provide normal glucose levels. Furthermore, therapies, such as islet transplantation or *in vitro* generation of beta-cells from stem/progenitor cells, still face a number of challenges, such as shortage of donors or inefficiency of insulin production, respectively. Recent research on genetic reprogramming of differentiated cells by the forced expression of key developmental factors has been successful in demonstrating that mature pancreatic cells still maintain a certain degree of plasticity [5, 24-26]. Indeed, fully differentiated cells are still capable of altering their phenotype and adopting a distinct functional identity.

As previously described, the transcription factors *Arx* and *Pax4* play major roles in the specification of the different endocrine cell subtypes: while Arx directs endocrine precursor cells towards the alpha-/PP-cell fates, Pax4 promotes the beta-/delta-cell lineages. In order to determine whether the expression of either of these factors was not only required, but also sufficient to generate a given mature cell subtype, each was misexpressed in the developing pancreas. The results obtained reveal that the misexpression of *Arx* in pancreatic or endocrine precursors leads to the nearly exclusive generation of alpha- and PP-cells, whereas almost no beta- or delta-cells are detected. Conversely, the misexpression of *Pax4* in developing pancreatic/endocrine cells results in the loss of alpha- and PP-cells concomitantly with a favoring of the beta-cell lineage [5]. These findings therefore demonstrate that these two factors are sufficient to favor specific lineages during pancreas morphogenesis. Most interesting was the discovery that the misexpression of *Pax4* in glucagon-expressing alpha-cells leads to a dramatic loss of alpha-cells accompanied by a drastic increase in the number of insulin-expressing beta-cells.

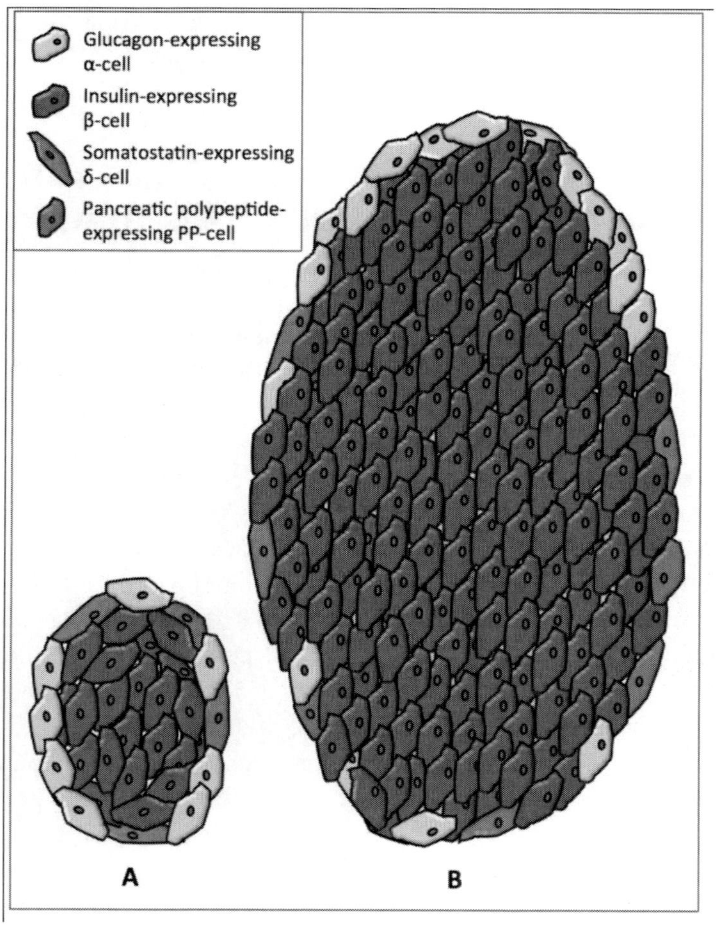

Figure 2. The misexpression of *Pax4* in glucagon-expressing alpha-cells leads to islet hypertrophy. Compared to age-matched wild-type controls (A), the islets of Langerhans of 12-week old animals misexpressing *Pax4* in glucagon-producing cells (B) are 4 to 8 times increased in size and literally filled with insulin-expressing beta-like cells.

Lineage tracing experiments indicated that *Pax4*-misexpressing alpha-cells are converted into cells exhibiting a beta-cell phenotype [5]. Further analysis revealed that alpha-cells are continuously regenerated and converted into beta-like cells. The source of such neo-generated cells is the duct-lining cells re-expressing the proendocrine factor Ngn3 (normally noticeable solely during pancreatic development). A comparable re-expression of *Ngn3* has previously only been reported after PDL (pancreatic duct ligation) - an injury model that generates tissue and cell damage [25]. The reconstitution of a developmental-like situation in animals ectopically expressing *Pax4* in alpha-cells could be caused by the glucagon shortage provoked by the alpha-to-beta-cell conversion or by an attempt to maintain the circulating insulin/glucagon ratio. Non-genetic lineage tracing experiments indicated that duct-lining cells continuously adopt a glucagon-expressing cell identity. However, these are subsequently converted into beta-like cells upon *Pax4* misexpression. Such cycle eventually results in a beta-like cell hyperplasia (Fig. 2B).

The continuously duct-generated progenitor cells adopting an alpha-cell identity, and their subsequent conversion into beta-like cells, may provide a valid explanation for the islet hypertrophy evident in animals misexpressing *Pax4* in alpha-cells (Fig. 2B). Importantly, when such animals are treated with streptozotocin - a chemical that is specifically toxic to beta-cells and used to chemically induce diabetes, a re-establishment of normal glycemia is noted, whereas all control mice are not capable of compensating their loss of insulin-expressing beta-cells and consequently suffer from lethal hyperglycemia. Interestingly, in this approach, younger animals are able to achieve euglycemia after streptozotocin treatment. Taken together, these results indicate that the misexpression of *Pax4* in glucagon-expressing alpha-cells not only induces a conversion of alpha-cells into functional beta-like cells, but also activates a cycle of alpha-cell-mediated beta-like cell regeneration.

Conclusion

New insights into the molecular pathways that regulate the generation of the pancreas and, in particular, the specification towards the different endocrine cell lineages, are of extreme importance in the search for a putative therapy for T1DM. Consequently, the potential clinical significance of Arx and Pax4 was recently shown to be crucial in directing endocrine precursor cells towards the different endocrine cell fates. Most important was the finding that the sole misexpression of the transcription factor Pax4 in alpha-cells induces their conversion into functional insulin-expressing beta-like cells that are capable of restoring euglycemia following complete beta-cell loss. Further research using microarray analyses may help to identify additional contributing participants in these signaling pathways. These factors may then represent putative targets for the design/screening of drugs allowing beta-like cell regeneration, thereby preventing cell de-differentiation into a pluripotent state and their subsequent re-differentiation. Should the approach presented in this chapter be translatable to the human pancreas, a new milestone in T1DM research would be achieved.

Acknowledgments

The authors are supported by the Juvenile Diabetes Research Foundation (17-2011-16, 2-2010-567, 26-2008-639), the European Research Council (StG-2011-28126), the INSERM AVENIR program, the INSERM, the Fondation pour la Recherche Médicale (DRC20091217179), the Agence Nationale de la Recherche (2009 GENO 105 01), Club Isatis, Mr. And Mrs. Olivier Dorato and the Foundation Schlumberger pour l'Education et la Recherche.

References

[1] Collombat P, Hecksher-Sorensen J, Serup P, Mansouri A. Specifying pancreatic endocrine cell fates. *Mech Dev.* 2006 Jul;123(7):501-12.

[2] Grapensparr L, Olerud J, Vasylovska S, Carlsson PO. The therapeutic role of endothelial progenitor cells in Type 1 diabetes mellitus. *Regen Med.* 2011 Sep;6(5):599-605.

[3] Deshpande AD, Harris-Hayes M, Schootman M. Epidemiology of diabetes and diabetes-related complications. *Phys Ther.* 2008 Nov;88(11):1254-64.

[4] Santamaria X, Massasa EE, Feng Y, Wolff E, Taylor HS. Derivation of Insulin Producing Cells From Human Endometrial Stromal Stem Cells and Use in the Treatment of Murine Diabetes. *Mol Ther.* 2011 Aug 30.

[5] Collombat P, Xu X, Ravassard P, Sosa-Pineda B, Dussaud S, Billestrup N, et al. The ectopic expression of Pax4 in the mouse pancreas converts

progenitor cells into alpha and subsequently beta cells. *Cell.* 2009 Aug 7;138(3):449-62.

[6] Pan FC, Wright C. Pancreas organogenesis: from bud to plexus to gland. *Dev Dyn.* [Research Support, N.I.H., ExtramuralReview]. 2011 Mar;240(3):530-65.

[7] Kloppe W. [Paul Langerhans (1847-1888) and his Berlin dissertation (1869)]. *Dtsch Med J.* 1969 Sep;20(18):581-3.

[8] Hauge-Evans AC, King AJ, Fairhall K, Persaud SJ, Jones PM. A role for islet somatostatin in mediating sympathetic regulation of glucagon secretion. *Islets.* 2010 Nov 1;2(6):341-4.

[9] Lee MG, Ahn W, Lee JA, Kim JY, Choi JY, Moe OW, et al. Coordination of pancreatic HCO3- secretion by protein-protein interaction between membrane transporters. *Jop.* 2001 Jul;2(4 Suppl):203-6.

[10] Wilson ME, Scheel D, German MS. Gene expression cascades in pancreatic development. *Mech Dev.* 2003 Jan;120(1):65-80.

[11] Guz Y, Montminy MR, Stein R, Leonard J, Gamer LW, Wright CV, et al. Expression of murine STF-1, a putative insulin gene transcription factor, in beta cells of pancreas, duodenal epithelium and pancreatic exocrine and endocrine progenitors during ontogeny. *Development.* 1995;121(1):11-8.

[12] Ohlsson H, Karlsson K, Edlund T. IPF1, a homeodomain-containing transactivator of the insulin gene. *Embo J.* 1993 Nov;12(11):4251-9.

[13] Gu G, Dubauskaite J, Melton DA. Direct evidence for the pancreatic lineage: NGN3+ cells are islet progenitors and are distinct from duct progenitors. *Development.* 2002 May;129(10):2447-57.

[14] Ahlgren U, Jonsson J, Edlund H. The morphogenesis of the pancreatic mesenchyme is uncoupled from that of the pancreatic epithelium in IPF1/PDX1-deficient mice. Development. 1996 May;122(5):1409-16.

[15] Offield MF, Jetton TL, Labosky PA, Ray M, Stein RW, Magnuson MA, et al. PDX-1 is required for pancreatic outgrowth and differentiation of the rostral duodenum. *Development.* 1996 Mar;122(3):983-95.

[16] Schwitzgebel VM, Scheel DW, Conners JR, Kalamaras J, Lee JE, Anderson DJ, et al. Expression of neurogenin3 reveals an islet cell precursor population in the pancreas. *Development.* 2000 Aug;127(16):3533-42.

[17] Gradwohl G, Dierich A, LeMeur M, Guillemot F. neurogenin3 is required for the development of the four endocrine cell lineages of the pancreas. *Proc Natl Acad Sci U S A.* 2000 Feb 15;97(4):1607-11.

[18] Grapin-Botton A, Majithia AR, Melton DA. Key events of pancreas formation are triggered in gut endoderm by ectopic expression of pancreatic regulatory genes. *Genes Dev.* 2001 Feb 15;15(4):444-54.
[19] Apelqvist A, Li H, Sommer L, Beatus P, Anderson DJ, Honjo T, et al. Notch signalling controls pancreatic cell differentiation. *Nature.* 1999 Aug 26;400(6747):877-81.
[20] Johansson KA, Dursun U, Jordan N, Gu G, Beermann F, Gradwohl G, et al. Temporal control of neurogenin3 activity in pancreas progenitors reveals competence windows for the generation of different endocrine cell types. *Dev Cell.* 2007 Mar;12(3):457-65.
[21] Sosa-Pineda B. The gene Pax4 is an essential regulator of pancreatic beta-cell development. *Mol Cells.* 2004 Dec 31;18(3):289-94.
[22] Blair HJ, Reed V, Gormally E, Wilson JB, Novak J, McInnes RR, et al. Positioning of five genes (CASK, ARX, SAT, IMAGE cDNAs 248928 and 253949) from the human X chromosome short arm with respect to evolutionary breakpoints on the mouse X chromosome. *Mamm Genome.* 2000 Aug;11(8):710-2.
[23] Collombat P, Hecksher-Sorensen J, Broccoli V, Krull J, Ponte I, Mundiger T, et al. The simultaneous loss of Arx and Pax4 genes promotes a somatostatin-producing cell fate specification at the expense of the alpha- and beta-cell lineages in the mouse endocrine pancreas. *Development.* 2005 Jul;132(13):2969-80.
[24] Zhou Q, Brown J, Kanarek A, Rajagopal J, Melton DA. In vivo reprogramming of adult pancreatic exocrine cells to beta-cells. *Nature.* 2008 Oct 2;455(7213):627-32.
[25] Xu X, D'Hoker J, Stange G, Bonne S, De Leu N, Xiao X, et al. Beta cells can be generated from endogenous progenitors in injured adult mouse pancreas. *Cell.* 2008 Jan 25;132(2):197-207.
[26] Thorel F, Nepote V, Avril I, Kohno K, Desgraz R, Chera S, et al. Conversion of adult pancreatic alpha-cells to beta-cells after extreme beta-cell loss. *Nature.* 2010 Apr 22;464(7292):1149-54.

In: Glucagon
Editor: Ahmed Al-Jebawi
ISBN: 978-1-62618-147-2

Chapter 8

Glucagon: Improving Knowledge and Education on the Antidote to Severe Hypoglycemia

B. Sommavilla*
University Hospital, Center for Prevention and Care,
Salzburg, Austria

Abstract

Although stringent glycemic control can protect against vascular complications in people with diabetes, it can also increase the risk of hypoglycemia.

Among other forms of administration, glucagon is available as an emergency kit to treat severe hypoglycemia, but while very effective, it is underutilized, suggesting the need for improved education surrounding its use.

This chapter details the role of glucagon in treating severe hypoglycemia and makes suggestions for how such education could be improved for people with diabetes and their caregivers to increase their confidence in administering glucagon in emergency situations.

* Email: b.sommavilla@gmx.at.

Introduction

People with type 1 or type2 diabetes must undertake stringent glycemic control in order to reduce and/or delay micro- and macro-vascular complications. Influential studies such as the Diabetes Control and Complications Trial (DCCT) and the United Kingdom Prospective Diabetes Study (UKPDS), among others, have presented conclusive evidence that implementation of strict glycemic thresholds delays the progression of diabetes complications [1-11], including microvascular complications, although the effects on macrovascular complications are less clear [12]. However, efforts to achieve strict metabolic control with intensive insulin therapy inevitably increase the risk of hypoglycaemia [6, 13-16]. In general, the guidelines for adults and children define an episode of hypoglycemia as being *mild, moderate,* or *severe*. *Mild* hypoglycemia is defined as a hypoglycemia event that the patient can self-treat, irrespective of symptom severity, or an asymptomatic blood glucose measurement of ≤3.9 mmol/l (70 mg/dl) [17-19]. With mild hypoglycemia, only autonomic symptoms are present, whereas both autonomic and neuroglycopenic symptoms are present with moderate hypoglycaemia [20]. *Severe* hypoglycemia is defined as a hypoglycemia event that requires external (third-party) assistance to treat [19]. The frequency of severe hypoglycemia is approximately one episode per patient per year in people with type1 diabetes, and only about 10–20% of this in people with type2 diabetes [21-23]. However, the risk of severe hypoglycemia in type 2 diabetes rises as the condition progresses and there is increased need for insulin; among patients with type2 diabetes, the incidence of severe hypoglycemia is highest among those treated with insulin [24]. Hypoglycemia is associated with significant morbidity, reductions in quality of life and, in some cases, can lead to death [24, 25]. Hypoglycemia symptoms can prove distressing and impair the daily functioning of people with diabetes, including work productivity [26-28]. Self-management of diabetes and quality of glycemic control are important tools that reduce the risk of hypoglycemia episodes, but can also be affected by fear of future severe hypoglycemia events [29-34]. Hypoglycemia characteristically results from a combination of excess therapeutic insulin (relative or absolute) and/or impaired physiological and behavioral responses to falling plasma glucose levels [24, 35-37]. Physiological defences against hypoglycemia include reduced insulin secretion, increased glucagon secretion and, in the absence of glucagon secretion, increased epinephrine secretion. These mechanisms are impaired in type 1 diabetes and long-standing type2 diabetes [35, 36, 38, 39]. Also, the threshold plasma glucose concentration that triggers counter-regulatory defences is lowered by previous exercise, recent prior hypoglycemia, or sleep, [35, 36, 38-

42] through mechanisms that are not yet fully defined but thought to involve the brain [24, 35, 36, 40]. Moreover, the compromised sympathetic neural response leads to hypoglycemia unawareness, such that the warning symptoms that normally alert the person with diabetes to ingest carbohydrates to counteract hypoglycemia are diminished or lost. Impaired awareness of hypoglycemia is a major risk factor, and makes it up to nine times more likely that a patient will have a severe hypoglycemia episode relative to those with normal awareness [43-45]. Impaired hypoglycemia awareness happens when the intensity of hypoglycemia symptoms reduces over time or the symptoms cease to develop. Approximately 25% of patients with type 1 diabetes have impaired hypoglycemia awareness, with a lower incidence in those with type 2 diabetes [43-47]. Because impaired hypoglycemia awareness makes the person with diabetes less able to detect the onset of hypoglycemia, the individual is less likely to start corrective action early enough and is more likely to go on to suffer severe episodes. In addition, impaired hypoglycemia awareness poses a high risk of self-injury (e.g., road traffic accidents, and accidents when working with machinery) or injuring others [48, 49]. As such, the effects and treatment of hypoglycemia place a significant burden on healthcare systems and society [50]. While regular monitoring of blood glucose levels and appropriate self-management are invaluable tools that help people to recognize the signs and symptoms of hypoglycemia (which is a particular issue in children with diabetes, as they find it more difficult to recognise the signs), and therefore, reduce the risk of hypoglycemia, there remains a need for precautionary measures to treat severe hypoglycemia episodes when they do occur. While intravenous (IV) dextrose can be administered by medical personnel to treat severe hypoglycemia, glucagon can be administered by non-medical personnel before the emergency services arrive, thereby allowing treatment to occur sooner, and possibly improve outcomes. Thus, glucagon is essential as a therapeutic strategy for managing episodes of severe hypoglycemia.

Glucagon as a Treatment for Severe Hypoglycemia

Efficacy, Safety and Tolerability

Under normal circumstances, it is secreted by the alpha cells of the pancreas, and maintains glucose production through glycogenolysis and gluconeogenesis in the liver [51, 52]. In emergency situations, when the blood sugar is severely low, it may not be possible for the body to produce a sufficient amount of glucagon quickly enough to counteract hypoglycemia.

Administration of exogenous (recombinant) glucagon is the first-line treatment for severe hypoglycemia given its role as the main counter-regulatory hormone to insulin. Recombinant glucagon is a polypeptide hormone that is identical to human glucagon that increases blood glucose levels and relaxes smooth muscle of the gastrointestinal tract. Glucagon signals to the liver to release its stored glucose into the blood, thereby returning the patient to a conscious state where further action can be taken. Unlike IV dextrose, which must be administered by a qualified healthcare professional, glucagon can be administered by subcutaneous (SC) or intramuscular (IM) injection by trained laypeople, like parents, friends, partners or caregivers. The pharmacokinetic and pharmacodynamic profiles of recombinant glucagon have been evaluated following IM and SC administration – peak plasma concentration is reached a few minutes after injection and the half-life is short (~8–18 minutes) [53, 54].

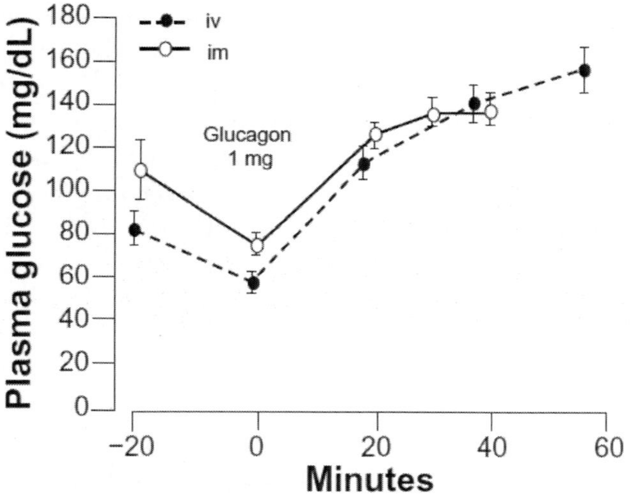

im = intramuscular; iv = intravenous.

Reprinted from Diabetes Res Clin Pract, 19, M Namba, T Hanafusa, N Kono, et al, Clinical evaluation of biosynthetic glucagon treatment for recovery from hypoglycemia developed in diabetic patients, 133–138, copyright (1993), with permission from Elsevier.

Figure 1. Changes in blood glucose levels before and after the administration of 1 mg biosynthetic glucagon.

An abundance of evidence supports the efficacy of glucagon for restoring blood glucose levels (Figure 1), and shows it is well-tolerated in treating severe hypoglycemia via IV, IM or SC injections [55, 56]. The side-effects of glucagon administration may include nausea and vomiting, particularly with doses >1 mg, but such reports are rare [54]. The risks associated with glucagon use are lower than those associated with severe hypoglycemia, and glucagon can lead to a faster recovery than would otherwise be the case if an ambulance is called and the patient waits for paramedics to start an IV line of dextrose.

It should be noted that glucagon is not recommended for IV infusion. Glucagon should be administered with caution to patients with the following conditions: chronic hypoglycemia, prolonged fasting, starvation, or adrenal insufficiency [54]. These conditions lead to reduced levels of releasable glucose in the liver, and glucagon requires adequate hepatic stores of glycogen to counteract hypoglycemia. Glucagon should also be used with caution in patients suspected of having insulinoma or glucagonoma [54].

Glucagon Emergency Kits

Glucagon is available as an emergency kit. The Glucagon Emergency Kit (Eli Lilly and Co., Indianapolis, IN, USA) and the GlucaGen® HypoKit (Novo Nordisk A/S, Bagsværd, Denmark) are the currently available glucagon kits administered by IM or SC injection (Figure 2). The glucagon kits include a brightly colored case, containing a vial of glucagon (which is freeze-dried into a powder to improve stability), a syringe prefilled with solvent, plus instructions for reconstitution and use. The design of the kits is such that they should be easy to find and use if an emergency arises.

Glucagon should only be administered when the person with diabetes is unconscious or unresponsive and unable to ingest oral carbohydrate. For adults and children over 25 kg in weight or children aged 6 to 8 years or above, a 1 mg dose of glucagon (reconstituted in 1 ml of sterile water) is the recommended dose, while for children below 25 kg in weight or younger than 6 to 8 years of age, a half dose (0.5 ml) is recommended [54].

In order to use the kit, the user must open the vial by breaking off the plastic cap, before injecting the solvent into the vial containing the glucagon powder, and then shaking the vial until the reconstituted solution is clear and has no particles. The user must then withdraw the reconstituted solution back into the syringe before immediately injecting it into the unconscious person's arm, leg or abdomen. Any unused drug should be disposed of. The unconscious person should immediately be rolled onto their side in case of vomiting during recovery. Because glucagon depletes liver glycogen stores, when the treated person awakens and is able to swallow, they should be given oral carbohydrates to restore glycogen stores and avoid secondary hypoglycemia. It is recommended that all people with diabetes who experience severe hypoglycemia undergo a medical evaluation. Recovery time depends on the blood sugar level prior to treatment and the duration of hypoglycaemia [57]. However, if there is no immediate response from the patient following glucagon injection (within 5–10 minutes), medical help should be sought and IV dextrose administered as soon as possible. Further information on the Glucagon Emergency Kit and the GlucaGen® Hypokit, can be found at visit http://pi.lilly.com/us/rglucagon-ppi.pdf and http://www.novonordisk .com/diabetes_care/pdf/11966_novo_glucagen_datasheets_11090621.pdf, respectively.

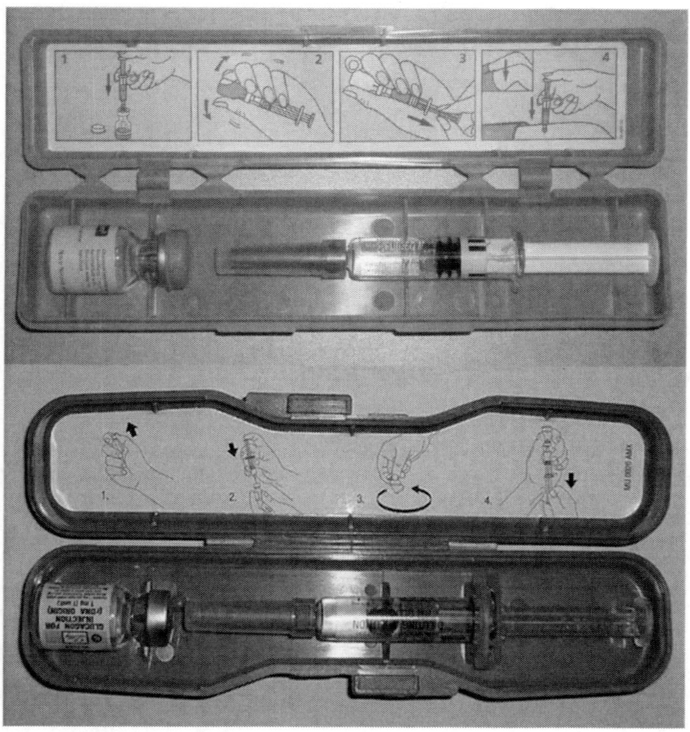

Figure 2. Top: The GlucaGen® HypoKit, easily identified by its orange box, and showing its components and the simple four-step instruction-by-picture label. Notice the orange plastic cap on the vial, which needs to be removed. Bottom: The Glucagon Emergency Kit, easily identified by its red box, and showing its components and the simple four-step instruction-by-picture label.

The American Diabetes Association (ADA) provides simple 'do/do not' advice for people assisting diabetes patients who experience an episode of severe hypoglycemia and lose consciousness. In concise form, the advice is: 'DO NOT inject insulin, provide food or fluids, or put hands in their mouth; DO inject glucagon and call for emergency help'. It is important not to put glucose tablets between the cheek and teeth when a person experiences an episode of severe hypoglycemia. In the 'Glucagon emergency administration training tool: a resource for school nurses and school personnel' developed by the New York State Department of Health in collaboration with the New York State Board of Nursing and the New York Statewide School Health Services Center [57], the recommendation for a student experiencing severe hypoglycemia states: 'It is virtually impossible to overdose on glucagon and therefore, always safer to treat in order to avoid the serious dangers of not treating.' The training tool recommends that the student experiencing severe hypoglycemia is treated on the spot and never left unattended. The school nurse/personnel should have another person follow the district policy for medical emergency care, or if they are alone, they should treat the student and follow the district policy for medical emergency care immediately. Furthermore, they should inform the emergency responders that they are treating a child with diabetes, in order for an appropriate emergency response team to answer the call.

Underuse of Glucagon

Despite the documented benefits, relative ease of administration, good safety and tolerability profile of glucagon administration for treating severe hypoglycemia, glucagon emergency kits remain underused [59-61]. There are several potential reasons for this: caregivers who are not used to administering by vial and syringe may worry that they will harm the patient, for example, relatives have been reported to be too anxious or unwilling to inject glucagon; [62] caregivers may not be aware of the causes (Table 1) and symptoms of hypoglycemia (e.g., sweating, dizziness, palpitations, tremor, hunger, restlessness, tingling in the hands, feet, lips, or tongue, light-headedness, headache, inability to concentrate, blurred vision, slurred speech, unsteady movement, personality changes, etc); or caregivers may not be aware that people with diabetes could have impaired hypoglycemia awareness such that the patient does not recognize the signs and symptoms of hypoglycaemia [60, 63]. In addition, healthcare professionals may not make their patients fully aware of the benefits of the emergency kits. Indeed, in a 1-year retrospective cohort study of three emergency departments, of the 436 hypoglycemia cases confirmed, recommendations to obtain glucagon emergency kits appeared in only 2% of written discharge instructions [64].

Table 1. Causes of hypoglycemia

Cause	Example
Incorrect insulin administration	Insulin taken in excess or at the wrong time relative to food intake and/or physical activity; incorrect type of insulin taken; insulin injected IM instead of SC
Insufficient exogenous carbohydrate	Delayed or missed meals or overnight fast
Decreased endogenous glucose production	Excess alcohol consumption
Increased utilization of carbohydrate/ depletion of hepatic glycogen stores	Exercise or weight loss
Increased insulin sensitivity	During the night, exercise or weight loss
Delayed gastric emptying	Conditions such as gastroparesis
Decreased insulin clearance	Conditions such as progressive renal failure

IM: intramuscularly; SC: subcutaneously.

Underuse could also be attributed to people being reluctant to administer glucagon in emergency situations (especially if they are not a family member or properly trained), due to fear of legal consequences if they do not correctly administer the drug. A related example is that, in some states in the USA, public school policy and some teachers' union contracts require that union teachers should not be allowed to deliver glucagon or even be trained in the administration of glucagon [59]. In the USA, there are several federal laws that protect the rights of students with disabilities to the same educational opportunities available to students without disabilities, including The Americans with Disabilities Act of 1992, The Individuals with Disabilities Education Act (IDEA) of 1991, and Section 504 of the Rehabilitation Act of 1973. Students with diabetes are covered by Section 504 and the Americans with Disabilities Act, and depending on services required, may also be covered by IDEA. These laws consider diabetes to be a disability, and serve to ensure optimal student safety at school and school-sponsored events. As it is illegal for schools to discriminate against children with disabilities, any school given federal funding or any facility considered open to the public must reasonably accommodate the needs of children with diabetes.

The school nurse is the most appropriate person to provide care for children with diabetes during school. However, many schools do not have a full-time nurse and one nurse may cover different schools. In addition, even a full-time school nurse may not be available throughout the school day, during extracurricular activities, or on field trips. In one study of school children with diabetes in the State of Virginia (where legislation requires non-medical school personnel to assist students with managing type 1 diabetes when school nurses are unavailable), among 185 parents whose children attended 153 different schools, 69% reported that a full-time school nurse was allocated to their child's school [65]. In other schools, part-time nurses assisted students with their diabetes management tasks, supplemented by non-medical personnel consisting of teachers, administrators, coaches, and cafeteria workers. The study found that among 185 students, 75% experienced a median of five hypoglycemia episodes per year. One severe event was reported, in which glucagon was administered appropriately by a part-time school nurse, and the student experienced no adverse effects related to treatment. Calculations based on a rate of one in 185 children experiencing severe hypoglycemia that required glucagon showed that, in a given school year, as many as 3% of children with diabetes could experience severe hypoglycemia. This study showed that children with type 1 diabetes can be safely cared for at school by medical and non-medical personnel. The widespread introduction of legislation enabling non-medical school personnel to assist students with managing their diabetes may increase their safety at school and improve overall glycemic control of such students. Such assistance at school may also help address the tendency for parents and students to aim for higher blood glucose levels in order to prevent hypoglycemia, which may lead to a higher risk of late complications [66].

Improving Education on the Use of Glucagon to Treat Severe Hypoglycemia

Underutilization of glucagon kits in patients at risk of severe hypoglycemia may partly be attributed to a lack of adequate education and training [59, 60, 67]. Distribution of information to patients and caregivers about diabetes self-management and hypoglycemia is increasingly carried out in the form of various educational materials or media including videos, brochures, and structured teaching. The need for improved education on hypoglycemia among friends, family and neighbours of patients was listed as a key issue in a recent focus group study [68]. Unsubstantiated fears about administering glucagon have also been reported among family members and friends of people with diabetes [59, 69], suggesting a need for improved education on the benefits of glucagon and the method of glucagon administration. Among one group of 136 parents of teenagers and young children with diabetes, 83% had received initial verbal education on glucagon administration by a diabetes educator; however, 69% showed handling difficulties and there were instances of the full doses of glucagon not being administered, suggesting that training on glucagon administration requires ‘hands on’ practice and follow-up assessment of skills [59].

Such training could include repeated practical training given to caregivers, including the practical steps that should be taken in an emergency situation, e.g., administer glucagon, place the patient in a stable position on their side, and then call for an ambulance. Family members and caregivers may also find it useful to repeatedly practice giving the injections with saline into a cushion (or a piece of fruit) so that they are more confident about injecting in a real emergency situation. Moreover, caregivers should be advised to know the location of the kit at all times, to store the kit at room temperature, away from temperature highs and lows, and not in direct sunlight (and glucagon should never be frozen), and to make sure that the expiration date has not passed. It may be helpful for patients or their carers to write the expiration date in their calendar as a reminder to purchase a new kit a week or so before the current one expires.

Once the liquid and powder have been mixed, any remaining liquid should not be stored, and should be discarded immediately after use. It would be prudent to have a spare kit available, possibly keeping one at home and using one for travelling purposes. It may also be useful to display a card at work/school or home in a noticeable position (such as near a telephone or by a computer screen), which indicates where the kit is stored so that others know of its location in an emergency and can quickly find it. If the patient or caregiver has any other questions, they should consult a doctor, nurse or pharmacist.

Professional diabetes organizations suggest that education of patients and caregivers should also include information on the main causes of hypoglycemia and self-management of diabetes [70]. It is important that all types of caregivers (including parents, partners, friends, sports instructors, [kindergarten] teachers, siblings and healthcare professionals) are aware of the causes and signs of hypoglycemia and of the risks associated with impaired hypoglycemia awareness. They should be informed about how to minimize/prevent hypoglycemia, but should also be educated on what to do once the patient has recovered from a severe hypoglycemia episode. For example, reasonable uniformity regarding diet, insulin administration and exercise may reduce the risk, as may frequent testing of blood glucose so that a change in insulin requirements can be foreseen. Another important piece of advice is to recommend that individuals with diabetes have an immediate source of glucose (for mild or moderate hypoglycemia episodes) and that a glucagon kit (in the case of severe hypoglycemia episodes) is available at all times.

Patients or their caregivers should inform their physician when hypoglycemia episodes occur, as it may be necessary to adjust the treatment regimen. Training on preparation for travelling, especially for a long journey, is essential, including ensuring that the glucagon kit is at hand. In addition, people with diabetes should be educated on how insulin intake should be adjusted after exercise to reduce the risk of hypoglycaemia [71].

The value of glucagon emergency kits may be particularly high for people with diabetes who have impaired hypoglycemia awareness (e.g., due to autonomic neuropathy [72], a long duration of diabetes or overly strict metabolic control), because if hypoglycemia is not recognized and treated quickly, severe episodes may ensue.

The ability to detect low blood glucose levels is a significant problem for children with type 1 diabetes and their parents, suggesting they may also particularly benefit from glucagon emergency kits, and better education and training to detect episodes of hypoglycemia in this population is needed [73]. To reflect this, the 2009 consensus guidelines developed by the International Society for Pediatric and Adolescent Diabetes (ISPAD) state that families of children and adolescents with diabetes should receive education about the situations in which hypoglycemia is more likely to occur, including: when the treatment regimen is altered; in younger children; during periods of frequent low blood glucose levels; when awareness of autonomic symptoms is reduced; or during sleep. Various professional organizations (such as the ADA and Centers for Disease Control and Prevention [CDC]) recommend that caregivers of children with diabetes make a plan to help them manage diabetes at school [74], including ensuring that school workers have a glucagon emergency kit and know how to use it. On a related note, healthcare professionals should be aware that a patient presenting with signs of drunkenness, such as sweating, becoming confused, aggressive or unusually loud, slurred speech, stumbling, losing their bearings, or passing out, may actually have diabetes and be experiencing hypoglycemia, which could ultimately become severe and require glucagon administration.

If a person with diabetes has in fact consumed alcohol, has adrenal insufficiency, or been fasting for a prolonged period of time, and experiences an episode of hypoglycemia, caregivers should be aware that glucagon has little or no effect because liver glycogen will be depleted. If the person does not respond to therapy, IV administration of dextrose by a healthcare provider at a hospital or other emergency services is warranted.

Conclusion

Strict glycemic control has been shown to delay the development of diabetes-related complications [1-11]. However, intensive insulin therapy to achieve tight glycemic control increases the risk of hypoglycemia,[6,13-16] which is associated with significant morbidity, a decline in quality of life and can be life-threatening.[24,25] In the case of a severe hypoglycemia episode where the patient requires third-party assistance to raise their blood glucose concentration, an abundance of evidence lends support to the consideration of glucagon as the treatment of choice in people with diabetes in non-hospital environments. Glucagon has a good safety profile, is well-tolerated, and can be administered by trained non-medical personnel, such as family members and caregivers. Injection of glucagon before arrival of the emergency services allows earlier restoration of blood glucose levels and can result in a more rapid recovery from severe hypoglycemia.

Further, glucagon has the potential to provide economic benefits by, for example, reducing the number of hospital admissions due to severe hypoglycemia and freeing up healthcare resources (Table 2) [75]. Glucagon treatment can also help patients and caregivers by reducing the anxiety and worry that accompany a severe hypoglycemia event. In spite of the benefits of glucagon administration in severe hypoglycemia, particularly for patients with hypoglycemia unawareness and pediatric patients, glucagon emergency kits are underutilized.[59-61] This may be explained in part by misconceived fears of glucagon administration, which have been reported among family members and friends of people with diabetes.[59] Such misconceptions are also seen in schools, clubs and other childcare facilitates leading to a further reluctance to inject glucagon.

This is particularly concerning given that there is an increasing number of people developing type 2 diabetes worldwide, as well as a shift to more intensive insulin regimens. Education and training may help to alleviate these fears, and such education could include the major causes and signs of hypoglycemia, the risks of impaired hypoglycemia awareness, the benefits of glucagon administration, and hands on training and repeated practice of glucagon injection. Better provision of information and continual education, combined with a greater awareness and more frequent use of glucagon may result not only in short-term benefits for managing severe hypoglycemia episodes, but long-term improvements due to better glycemic control and fewer diabetes-related late complications.

Table 2. Costs associated with severe hypoglycemia in three European countries (based on 2007 prices). Adapted from J Med Econ, 12, M Hammer, M Lammert, SM Mejias, et al, Costs of managing severe hypoglycaemia in three European countries, 281–290, copyright (2009), with permission from Informa Healthcare Communications

	Germany (€)	Spain (€)	UK (UK £)
HCP consultations			
PCP office visit	7.40	15.10	32.42
PCP phone call	7.40	10.11	28.23
Diabetes specialist	19.40	52.35	112.93
Emergency HCP access costs			
Ambulance	281.74	330.91	187.17
Emergency physician	318.86	65.52	72.15
Taxi transfer	21.48	10.55	10.00
Private car transfer	4.10	5.00	2.60
Emergency call (treatment guidance)	7.40	n/a	10.46
Hotline call (emergency)	7.40	n/a	10.46
Hospital costs			
Emergency room	140.61	786.91	73.00
Intensive care unit	n/a	1,264.61	1,657.37
General ward	n/a	405.90	502.96
Hospitalization tariff	2,517.00	n/a	612.76
Treatments			
Sugar (estimate)	0.09	0.05	0.01
Sweet drinks (estimate)	0.50	0.45	0.35
Sweets, chocolate (estimate)	0.50	0.93	0.28
Oral glucose gel/paste	0.75	5.85	3.65
Glucagon (im)	27.74	21.40	11.52
Glucose (iv)	27.74	1.86	2.55
Training and other costs			
Education	7.40	29.66	32.42
Dietary counseling	7.40	29.66	32.42
Hypoglycemia awareness training	7.40	29.66	32.42
Psychological counseling	7.40	29.66	32.42
Leaflets, books	2.21	2.21	1.50
Blood glucose test (patient self-test)	0.50	0.89	0.29
Salary per hour	20.74	10.67	11.92

Abbreviations: HCP = healthcare professional; PCP = primary care physician; n/a = not applicable; im = intramuscular; iv = intravenous.

Conflict of Interest Statement

The author has no conflicts of interest to declare.

Acknowledgments

This chapter was sponsored by Novo Nordisk A/S (Bagsværd, Denmark). Editorial support was provided by ESP Bioscience (Crowthorne, UK) funded by Novo Nordisk A/S (Bagsværd, Denmark).

References

[1] Davila EP, Florez H, Trepka MJ, et al. Strict glycemic control and mortality risk among US adults with type 2 diabetes. *J Diabetes Complications.* Sep-Oct 2011;25(5):289-291.

[2] Franklin VL, Khan F, Kennedy G, Belch JJ, Greene SA. Intensive insulin therapy improves endothelial function and microvascular reactivity in young people with type 1 diabetes. *Diabetologia.* Feb 2008;51(2):353-360.

[3] Reichard P, Nilsson BY, Rosenqvist U. The effect of long-term intensified insulin treatment on the development of microvascular complications of diabetes mellitus. *N Engl J Med.* Jul 29 1993;329(5):304-309.

[4] Skyler JS, Bergenstal R, Bonow RO, et al. Intensive glycemic control and the prevention of cardiovascular events: implications of the ACCORD, ADVANCE, and VA diabetes trials: a position statement of the American Diabetes Association and a scientific statement of the American College of Cardiology Foundation and the American Heart Association. *Diabetes Care.* Jan 2009;32(1):187-192.

[5] Stratton IM, Adler AI, Neil HA, et al. Association of glycaemia with macrovascular and microvascular complications of type 2 diabetes (UKPDS 35): prospective observational study. *BMJ.* Aug 12 2000;321(7258):405-412.

[6] Wang PH, Lau J, Chalmers TC. Meta-analysis of effects of intensive blood-glucose control on late complications of type I diabetes. *Lancet.* May 22 1993;341(8856):1306-1309.

[7] UK Prospective Diabetes Study (UKPDS) Group. Intensive blood-glucose control with sulphonylureas or insulin compared with conventional

treatment and risk of complications in patients with type 2 diabetes (UKPDS 33). UK Prospective Diabetes Study (UKPDS) Group. *Lancet.* Sep 12 1998;352(9131):837-853.

[8] Nathan DM, Zinman B, Cleary PA, et al. Modern-day clinical course of type 1 diabetes mellitus after 30 years' duration: the diabetes control and complications trial/epidemiology of diabetes interventions and complications and Pittsburgh epidemiology of diabetes complications experience (1983-2005). *Arch Intern Med.* Jul 27 2009;169(14):1307-1316.

[9] The absence of a glycemic threshold for the development of long-term complications: the perspective of the Diabetes Control and Complications Trial. *Diabetes.* Oct 1996;45(10):1289-1298.

[10] Hypoglycemia in the Diabetes Control and Complications Trial. The Diabetes Control and Complications Trial Research Group. *Diabetes.* Feb 1997;46(2):271-286.

[11] de Boer IH, Sun W, Cleary PA, et al. Intensive diabetes therapy and glomerular filtration rate in type 1 diabetes. *N Engl J Med.* Dec 22 2011;365(25):2366-2376.

[12] Brown A, Reynolds LR, Bruemmer D. Intensive glycemic control and cardiovascular disease: an update. *Nat Rev Cardiol.* Jul 2010;7(7):369-375.

[13] Buehler AM, Cavalcanti AB, Berwanger O, et al. Effect of Tight Blood Glucose Control Versus Conventional Control in Patients with Type 2 Diabetes Mellitus: A Systematic Review with Meta-Analysis of Randomized Controlled Trials. *Cardiovasc Ther.* Dec 29 2011.

[14] Hemmingsen B, Lund SS, Gluud C, et al. Targeting intensive glycaemic control versus targeting conventional glycaemic control for type 2 diabetes mellitus. *Cochrane Database Syst Rev.* 2011(6):CD008143.

[15] Hemmingsen B, Lund SS, Gluud C, et al. Intensive glycaemic control for patients with type 2 diabetes: systematic review with meta-analysis and trial sequential analysis of randomised clinical trials. *BMJ.* 2011;343:d6898.

[16] Shalitin S, Phillip M. Hypoglycemia in type 1 diabetes: a still unresolved problem in the era of insulin analogs and pump therapy. *Diabetes Care.* Feb 2008;31 Suppl 2:S121-124.

[17] Ababou MR. Hypoglycemia in the Diabetes Control and Complications Trial. The Diabetes Control and Complications Trial Research Group. *Diabetes.* Feb 1997;46(2):271-286.

[18] Clarke W, Jones T, Rewers A, Dunger D, Klingensmith GJ. Assessment and management of hypoglycemia in children and adolescents with diabetes. *Pediatr Diabetes.* Sep 2009;10 Suppl 12:134-145.

[19] Strachan MWJ. Frequency, causes, and risk factors for hypoglycaemia in type 1 diabetes. In: Frier BM, Fisher BM, eds. *Hypoglycaemia in clinical diabetes.* Chichester: John Wiley; 2007:49-81.

[20] Committee CDACPGE. Canadian Diabetes Association 2008 Clinical Practice Guidelines for the Prevention and Management of Diabetes in Canada. *Canadian Journal of Diabetes.* 2008;32(Suppl 1):S1-S201.

[21] Leese GP, Wang J, Broomhall J, et al. Frequency of severe hypoglycemia requiring emergency treatment in type 1 and type 2 diabetes: a population-based study of health service resource use. *Diabetes Care.* Apr 2003;26(4):1176-1180.

[22] MacLeod KM, Hepburn DA, Frier BM. Frequency and morbidity of severe hypoglycaemia in insulin-treated diabetic patients. *Diabet Med.* Apr 1993;10(3):238-245.

[23] Reichard P, Berglund B, Britz A, Cars I, Nilsson BY, Rosenqvist U. Intensified conventional insulin treatment retards the microvascular complications of insulin-dependent diabetes mellitus (IDDM): the Stockholm Diabetes Intervention Study (SDIS) after 5 years. *J Intern Med.* Aug 1991;230(2):101-108.

[24] Cryer PE, Davis SN, Shamoon H. Hypoglycemia in diabetes. *Diabetes Care.* Jun 2003;26(6):1902-1912.

[25] Zoungas S, Patel A, Chalmers J, et al. Severe hypoglycemia and risks of vascular events and death. *N Engl J Med.* Oct 7 2010;363(15):1410-1418.

[26] Brod M, Christensen T, Thomsen TL, Bushnell DM. The impact of non-severe hypoglycemic events on work productivity and diabetes management. *Value Health.* Jul-Aug 2011;14(5):665-671.

[27] Wredling R, Levander S, Adamson U, Lins PE. Permanent neuropsychological impairment after recurrent episodes of severe hypoglycaemia in man. *Diabetologia.* Mar 1990;33(3):152-157.

[28] Defining and reporting hypoglycemia in diabetes: a report from the American Diabetes Association Workgroup on Hypoglycemia. *Diabetes Care.* May 2005;28(5):1245-1249.

[29] Banck-Petersen P, Larsen T, Perdersen-Bjergaard U, Bie-Olsen L, Høi-Hansen T, Thorsteinsson B. Concerns about hypoglycamia and late complications in patients with insulin-treated diabetes. *Eur Diabetes Nursing.* 2007;4:113-118.

[30] Brod M, Kongso JH, Lessard S, Christensen TL. Psychological insulin resistance: patient beliefs and implications for diabetes management. *Qual Life Res.* Feb 2009;18(1):23-32.

[31] Currie CJ, Poole CD, Woehl A, et al. The health-related utility and health-related quality of life of hospital-treated subjects with type 1 or type 2 diabetes with particular reference to differing severity of peripheral neuropathy. *Diabetologia.* Oct 2006;49(10):2272-2280.

[32] Davis RE, Morrissey M, Peters JR, Wittrup-Jensen K, Kennedy-Martin T, Currie CJ. Impact of hypoglycaemia on quality of life and productivity in type 1 and type 2 diabetes. *Curr Med Res Opin.* Sep 2005;21(9):1477-1483.

[33] Haugstvedt A, Wentzel-Larsen T, Graue M, Sovik O, Rokne B. Fear of hypoglycaemia in mothers and fathers of children with Type 1 diabetes is associated with poor glycaemic control and parental emotional distress: a population-based study. *Diabet Med.* Jan 2010;27(1):72-78.

[34] Wild D, von Maltzahn R, Brohan E, Christensen T, Clauson P, Gonder-Frederick L. A critical review of the literature on fear of hypoglycemia in diabetes: Implications for diabetes management and patient education. *Patient Educ Couns.* Sep 2007;68(1):10-15.

[35] Cryer PE. Glucose homeostasis and hypoglycemia. In: Kronenberg H, Melmed S, Polonsky K, Larsen P, eds. *Williams textbook of endocrinology.* 11th ed. Philadelphia: Saunders, an imprint of Elsevier, Inc.; 2008:1503-1533.

[36] Cryer PE. Diverse causes of hypoglycemia-associated autonomic failure in diabetes. *N Engl J Med.* May 27 2004;350(22):2272-2279.

[37] Cryer PE. Hypoglycaemia: the limiting factor in the glycaemic management of Type I and Type II diabetes. *Diabetologia.* Jul 2002;45(7):937-948.

[38] Dagogo-Jack SE, Craft S, Cryer PE. Hypoglycemia-associated autonomic failure in insulin-dependent diabetes mellitus. Recent antecedent hypoglycemia reduces autonomic responses to, symptoms of, and defense against subsequent hypoglycemia. *J Clin Invest.* Mar 1993;91(3):819-828.

[39] Segel SA, Paramore DS, Cryer PE. Hypoglycemia-associated autonomic failure in advanced type 2 diabetes. *Diabetes.* Mar 2002;51(3):724-733.

[40] Cryer PE. Mechanisms of sympathoadrenal failure and hypoglycemia in diabetes. *J Clin Invest.* Jun 2006;116(6):1470-1473.

[41] Cryer PE. Mechanisms of hypoglycemia-associated autonomic failure and its component syndromes in diabetes. *Diabetes.* Dec 2005;54(12):3592-3601.

[42] Heller SR, Cryer PE. Reduced neuroendocrine and symptomatic responses to subsequent hypoglycemia after 1 episode of hypoglycemia in nondiabetic humans. *Diabetes.* Feb 1991;40(2):223-226.

[43] Clarke WL, Cox DJ, Gonder-Frederick LA, Julian D, Schlundt D, Polonsky W. Reduced awareness of hypoglycemia in adults with IDDM. A

prospective study of hypoglycemic frequency and associated symptoms. *Diabetes Care.* Apr 1995;18(4):517-522.

[44] Gold AE, MacLeod KM, Frier BM. Frequency of severe hypoglycemia in patients with type I diabetes with impaired awareness of hypoglycemia. *Diabetes Care.* Jul 1994;17(7):697-703.

[45] Hepburn DA, Patrick AW, Eadington DW, Ewing DJ, Frier BM. Unawareness of hypoglycaemia in insulin-treated diabetic patients: prevalence and relationship to autonomic neuropathy. *Diabet Med.* Sep-Oct 1990;7(8):711-717.

[46] Henderson JN, Allen KV, Deary IJ, Frier BM. Hypoglycaemia in insulin-treated Type 2 diabetes: frequency, symptoms and impaired awareness. *Diabet Med.* Dec 2003;20(12):1016-1021.

[47] Hepburn DA, Patrick AW, Brash HM, Thomson I, Frier BM. Hypoglycaemia unawareness in type 1 diabetes: a lower plasma glucose is required to stimulate sympatho-adrenal activation. *Diabet Med.* Dec 1991;8(10):934-945.

[48] Cox DJ, Kovatchev B, Vandecar K, Gonder-Frederick L, Ritterband L, Clarke W. Hypoglycemia preceding fatal car collisions. *Diabetes Care.* Feb 2006;29(2):467-468.

[49] Stork AD. [Hypoglycaemia unawareness: judge driving ability prudently]. *Ned Tijdschr Geneeskd.* 2011;155(48):A4144.

[50] Lammert M, Hammer M, Frier BM. Management of severe hypoglycaemia: cultural similarities, differences and resource consumption in three European countries. *J Med Econ.* 2009;12(4):269-280.

[51] Fonjallaz P, Loumaye E. Glucagon rDNA origin (GlucaGen) and recobinant LH. *J Biotechnol.* 2000;79:185-189.

[52] Unger RH, Orci L. Glucagon. In: Kahn CR, Weir GC, eds. *Joslin's Diabetes Mellitus.* Philadelphia, PA: Lea & Febiger; 1994:1306-1309.

[53] Graf CJ, Woodworth JR, Seger ME, Holcombe JH, Bowsher RR, Lynch R. Pharmacokinetic and glucodynamic comparisons of recombinant and animal-source glucagon after IV, IM, and SC injection in healthy volunteers. *J Pharm Sci.* Oct 1999;88(10):991-995.

[54] GlucaGen HypoKit [prescribing information]. *Novo Nordisk, A/S.* Denmark 2008.

[55] Carstens S, Sprehn M. Prehospital treatment of severe hypoglycaemia: a comparison of intramuscular glucagon and intravenous glucose. *Prehosp Disaster Med.* Apr-Dec 1998;13(2-4):44-50.

[56] Namba M, Hanafusa T, Kono N, Tarui S. Clinical evaluation of biosynthetic glucagon treatment for recovery from hypoglycemia developed in diabetic

patients. The GL-G Hypoglycemia Study Group. *Diabetes Res Clin Pract.* Feb 1993;19(2):133-138.

[57] The New York State Department of Health. Glucagon emergency administration training tool: a resource for school nurses and school personnel. *http://www.health.ny.gov/diseases/conditions/diabetes/docs/ glucagon_training.pdf.* Accessed July, 2012.

[58] Living with Diabetes. *http://www.diabetes.org/living-with-diabetes/treatment-and-care/blood-glucose-control/hypoglycemia-low-blood.html.* Accessed February, 2012.

[59] Harris G, Diment A, Sulway M, Wilkinson M. Glucagon administration—underevaluated and undertaught. *Pract Diabetes Int.* 2001;18:22-25.

[60] Kedia N. Treatment of severe diabetic hypoglycemia with glucagon: an underutilized therapeutic approach. *Diabetes, metabolic syndrome and obesity: targets and therapy.* 2011;4:337-346.

[61] Pearson T. Glucagon as a Treatment of Severe Hypoglycemia - Safe and Efficacious but Underutilized. *Diabet Educ.* 2008;34(1):128-134.

[62] Muhlhauser I, Berger M, Sonnenberg G, et al. Incidence and management of severe hypoglycemia in 434 adults with insulin-dependent diabetes mellitus. *Diabetes Care.* May-Jun 1985;8(3):268-273.

[63] Schopman JE, Geddes J, Frier BM. Frequency of symptomatic and asymptomatic hypoglycaemia in Type 1 diabetes: effect of impaired awareness of hypoglycaemia. *Diabet Med.* Mar 2011;28(3):352-355.

[64] Ginde AA, Pallin DJ, Camargo CA, Jr. Hospitalization and discharge education of emergency department patients with hypoglycemia. *Diabetes Educ.* Jul-Aug 2008;34(4):683-691.

[65] Hellems MA, Clarke WL. Safe at school: a Virginia experience. *Diabetes Care.* Jun 2007;30(6):1396-1398.

[66] Wennick A, Lundqvist A, Hallstrom I. Everyday experience of families three years after diagnosis of type 1 diabetes in children: a research paper. *J Pediatr Nurs.* Jun 2009;24(3):222-230.

[67] Yardley D, Lyddall A, Richardson J, et al. Glucagon injection for type 1 diabetes in children. *Nurs Child Young People.* Nov 2011;23(9):12-18.

[68] Sutton L, Chapman-Novakofski K. Hypoglycemia Education Needs. *Qual Health Res.* Mar 10 2011.

[69] Patton SR, Dolan LM, Henry R, Powers SW. Parental fear of hypoglycemia: young children treated with continuous subcutaneous insulin infusion. *Pediatr Diabetes.* Dec 2007;8(6):362-368.

[70] DCE Patient Education Handouts. *http://www.dce.org/publications/ education-handouts/.* Accessed April, 2011.

[71] Younk LM, Mikeladze M, Tate D, Davis SN. Exercise-related hypoglycemia in diabetes mellitus. *Expert Rev Endocrinol Metab.* Jan 1 2011;6(1):93-108.

[72] Meyer C, Hering BJ, Grossmann R, et al. Improved glucose counterregulation and autonomic symptoms after intraportal islet transplants alone in patients with long-standing type I diabetes mellitus. *Transplantation.* Jul 27 1998;66(2):233-240.

[73] Gonder-Frederick L, Zrebiec J, Bauchowitz A, et al. Detection of hypoglycemia by children with type 1 diabetes 6 to 11 years of age and their parents: a field study. *Pediatrics.* Mar 2008;121(3):e489-495.

[74] Klingensmith G, Kaufman F, Schatz D, Clarke W. Care of children with diabetes in the school and day care setting. *Diabetes Care.* Jan 2003;26 Suppl 1:S131-135.

[75] Hammer M, Lammert M, Mejias SM, Kern W, Frier BM. Costs of managing severe hypoglycaemia in three European countries. *J Med Econ.* 2009;12(4):281-290.

Index

A

B

C

D

E

Q

R

T

U

V

W

X

Y

Z